Particle Swarm Optimizer and Multi-Objective Optimization

Feng Pan · Qi Gao · Xiao-xue Feng · Wei-xing Li

Particle Swarm Optimizer and Multi-Objective Optimization

北京理工大学出版社
BEIJING INSTITUTE OF TECHNOLOGY PRESS

Springer

Feng Pan
School of Automation
Beijing Institute of Technology
Beijing, China

Qi Gao
School of Automation
Beijing Institute of Technology
Beijing, China

Xiao-xue Feng
School of Automation
Beijing Institute of Technology
Beijing, China

Wei-xing Li
School of Automation
Beijing Institute of Technology
Beijing, China

ISBN 978-981-95-3380-0   ISBN 978-981-95-3381-7   (eBook)
https://doi.org/10.1007/978-981-95-3381-7

Jointly published with Beijing Institute of Technology Press
The print edition is not for sale in China (Mainland). Customers from China (Mainland) please order the print book from: Beijing Institute of Technology Press.

This Springer imprint is published by the registered company Springer Nature Singapore Pte Ltd.
The registered company address is: 152 Beach Road, #21-01/04 Gateway East, Singapore 189721, Singapore

If disposing of this product, please recycle the paper.

Content Overview

Inspired by the mechanisms of biological evolution, researchers have proposed various novel methods to address complex problems, such as genetic algorithms, evolutionary programming, and evolutionary strategies. As an evolutionary computation technique, swarm intelligence algorithms have captured increasing attention from scholars, demonstrating profound connections with artificial life, particularly evolutionary strategies and genetic algorithms. Grounded in the collective behaviors of social animals and artificial life theory, swarm intelligence investigates the intrinsic principles of group behavior and designs innovative problem-solving methodologies based on these mechanisms. The fascinating aspect of social animals lies in their simplicity at the individual level, yet their collaborative efforts can yield highly complex (intelligent) behaviors.

This book presents a comprehensive exploration of Particle Swarm Optimization (PSO), including algorithmic foundations, theoretical analyses, and applications to both single-objective and multi-objective optimization. The book is organized into nine chapters. Chapter 1 serves as an introduction, offering a brief overview of optimization problems, swarm intelligence algorithms, and the current research landscape on PSO.

Chapter 2 provides a detailed exposition of the PSO algorithm. It offers a formal description, summarizes the common mathematical models, and discusses the algorithm's topological structures and evaluation metrics. The chapter concludes with a discussion of the diversity assessment metrics specific to PSO.

Chapters 3–6 analyze the characteristics of the PSO algorithm from a theoretical perspective. Beginning with the standard PSO model, these chapters examine and summarize issues such as premature convergence, as well as the properties of Gbest (global best), Pbest (personal best) and the standard PSO model itself. The stability criterion of PSO is analyzed as a dynamic time-varying system, supported by experimental validation of stability principles. Furthermore, the Markov properties of the standard PSO algorithm are examined according to Markov chain theory in stochastic systems. The impacts of inertia weights and acceleration factors on convergence are systematically discussed, followed by a summary of the sampling distribution and particle trajectories in the standard PSO.

Chapter 7 reviews improved PSO variants for single-objective optimization, including topology-based PSO enhancements, mathematical model improvements, hybrid PSO algorithms, and multi-swarm PSO algorithms. The chapter also outlines the underlying ideas behind these improvements and the corresponding algorithmic workflows.

Chapters 8 and 9 focus on multi-objective optimization with PSO. After introducing the mathematical formulation and classification of multi-objective optimization problems, Multi-Objective PSO (MOPSO) and its categorization based on inherent characteristics are presented. Two widely used MOPSO variants, CMOPSO and MOCLPSO, are described, refined, and experimentally evaluated. Additionally, a distance-based improved MOPSO algorithm (DISMOPSO) is proposed and benchmarked. The final section introduces a novel Cooperative multi-swarm MOPSO algorithm (CMPSO) and provides comparative performance analyses.

The book presents innovative content with a broad scope, offering clear and accessible explanations. It emphasizes the integration of theory and practical experiments, making it suitable not only for beginners but also as a reference for graduate and advanced undergraduate students in electronics, automation, computer science, information science, and related fields at universities and research institutes. Researchers and engineers in academia or industry may also find it valuable.

Given the author's limitations, some shortcomings and errors are inevitable. Constructive feedback from peers and readers is sincerely welcomed.

Introduction

Optimization technology, a computer-based application rooted in mathematics, is designed to identify optimal or satisfactory solutions to various engineering challenges. Currently, optimization techniques have been widely applied in numerous fields, including industry, agriculture, defense, engineering, transportation, finance, chemical processing, energy, communications and so on. For instance, in engineering design, optimization facilitates the selection of parameters to meet design specifications while achieving optimal performance. In resource allocation, it enables the effective distribution of limited resources to fulfill essential requirements and maximize economic returns. In summary, optimization technology enhances economic efficiency, improves system performance, and reduces energy consumption, thereby demonstrating broad prospects for national economic development.

With the rapid advancement of the modern economy, optimization has become an urgent priority. However, problems derived from rigorous mechanistic models often exhibit characteristics such as large-scale complexity, strong constraints, nonlinearity, multiple extrema, and multi-objective criteria—features that classical optimization algorithms struggle to address. Consequently, the development of optimization algorithms capable of handling large-scale problems has emerged as a critical focus in computational research.

The technological progress of the 20th century stands as a milestone in humanity's evolutionary journey. The explosive growth of information technology has propelled unprecedented scientific advancements in the new millennium. In 1834, *Charles Babbage*, the Lucasian Professor at the University of Cambridge, proposed the concept of the Analytical Engine, laying the groundwork for intelligent machinery. A century later, the famous "Turing Test" catalyzed the rapid development of artificial intelligence (AI). Since the 1980s, traditional AI, which relied on symbolic systems to simulate human intelligence, has faced growing challenges in knowledge representation, pattern processing, and combination explosion problem. This spurred the emergence of heuristic models inspired by "mimetic" or "bionic" concepts, such as neural networks (NN), evolutionary computation (EC), and fuzzy logic (FL). These approaches, enriched AI research, provided novel solutions to nonlinear and complex optimization problems. By 1994, computational intelligence (CI)—centered on EC,

FL, and NN—had coalesced into a distinct field, as marked by the inaugural World Congress on Computational Intelligence in Orlando, USA.

Computational Intelligence (CI) primarily focuses on developing advanced information processing techniques. However, a unified definition has not yet been established. *Bezdek* [1] rigorously distinguishes CI as intelligence dependent on numerical data, contrasting it with knowledge-based AI. Generally, any approach that uses computational methods to achieve intelligent behavior falls under CI. The primary goal of CI is to address real-world problems related to decision-making, modeling, and control, which are often poorly defined and require human intervention.

When search spaces grow too vast for exhaustive or blind search and heuristic knowledge proves inadequate, swarm-based optimization offers an effective solution. Swarm intelligence exemplifies this principle: simple agents collaborate to exhibit sophisticated behaviors, where collective wisdom surpasses individual capabilities. Current swarm intelligence methodologies in CI intersect with multiple disciplines, including AI, computer science, sociology, economics, ecology, organizational management, and philosophy. These interdisciplinary connections have led to the creation of new research domains, advancing broader scientific progress [2]. Particularly in complexity science, swarm intelligence offers a universal framework for solving intricate optimization problems that are difficult to define precisely in complex systems.

Social organisms, such as bees and ants, leverage collective effort for foraging, defense, and nesting. This emergent "intelligence" is termed swarm intelligence. By studying the principles underlying these collaborative behaviors, researchers have developed novel algorithms within the realm of computational intelligence, specifically in swarm intelligence. Studies of human intelligence reveal that most activities involve social groups and that solving complex problems often requires collaboration among specialists. Cooperation is one of the primary manifestations of human intelligence. While evolutionary computation mimics biological evolution, swarm intelligence algorithms simulate the social systems of gregarious organisms. In the field of CI, numerous algorithms based on swarm intelligence have been developed, such as ant colony optimization (ACO) and Particle Swarm Optimization (PSO).

Proposed by *James Kennedy* and *Russell Eberhart* in 1995, PSO is a parallel stochastic optimization algorithm inspired by the foraging behavior of bird flocks. In this model, birds search randomly for food in an unknown environment, dynamically following the flock's leader. Sociobiologist *E. O. Wilson* observed that, at least theoretically, individual members of a group benefit from the discoveries and experiences of all other members during the search for food. When food sources are unpredictably distributed, this cooperative advantage becomes decisive, far outweighing competitive drawbacks. PSO adapts this biological paradigm to solve optimization problems by leveraging social information sharing among agents.

Contents

Chapter 1
Preface

1.1 Optimization Problem

An optimization problem involves finding a set of parameter values that satisfy certain constraints and fulfill specific optimality criteria, thereby maximizing or minimizing particular performance metrics of a system. The importance of optimization problems is evident from their wide-range applications in industry, society, economics, and management and so on.

Optimization problems can be categorized into various types based on the nature of their objective functions, constraint functions, and the characteristics of their optimization variables. Each type often has its own specialized solution methods due to its unique characteristics.

In general terms, an optimization problem can be expressed as:

$$\begin{aligned} \min / \max \quad & \{y = f(x)\} \\ s.t. \quad & x \in S = \{x | g_i(x) \le 0, i = 1, 2, \cdots, m\} \end{aligned} \tag{1.1}$$

where $f(x)$ represents the objective function, $g_i(x)$ denote the constraint functions, and m is the number of constraints. S is the feasible region, while $x \in D$ represents the the vector of optimization variables within their domain D.

When both $f(x)$ and the constraint functions $g_i(x)$ are linear functions and $x \ge 0$, the problem becomes a linear programming problem. Well-established solution methods include the Simplex method and the Karmarkar algorithm, etc.

If at least one of these functions (objective or constraints) is nonlinear, the problem is classified as a nonlinear programming problem. Nonlinear programming is relatively complex, offering a variety of solution methods; however, to date, no single method has proven universally effective for all types of problems.

When the feasible space defined by $g_i(x) \le 0 \, (i = 1, 2, \cdots, m)$ spans the entire D-dimensional Euclidean space R^n, the problem reduces to an unconstrained optimization problem:

© Beijing Institute of Technology Press 2025
F. Pan et al., *Particle Swarm Optimizer and Multi-Objective Optimization*,
https://doi.org/10.1007/978-981-95-3381-7_1

$$\min / \max\{y = f(x)\}$$
$$s.t. \quad x \in S \subset R^n \tag{1.2}$$

Nonlinear programming problems (whether unconstrained or constrained) are challenging to solve due to the inherent nonlinearity of the functions, especially when multiple local optima exist. Common methods for nonlinear optimization are highly sensitive to initial values. In other words, typical methods for both constrained and unconstrained nonlinear optimization tend to converge to approximate local extrema rather than the global optimum.

The problem described above is a single-objective optimization problem. However in real-world economic, industrial, and engineering applications, decision-making often involves multiple, **potentially conflicting**, objectives, leading to what are known as Multi-Objective optimization Problems (MOP). When considering k objectives, these problems can be described as follows:

$$\min / \max\{\boldsymbol{F}(\boldsymbol{x}) = (f_1(\boldsymbol{x}), f_2(\boldsymbol{x}), f_3(\boldsymbol{x}) \ldots f_m(\boldsymbol{x}))\}$$
$$s.t. \quad x \in S = \{x | g_i(x) \le 0, i = 1, 2, \cdots, m\} \tag{1.3}$$

where $F(x)$ represents the vector of optimization objectives, $g_i(x)$ are the constraint functions, and x is the decision variable vector. For multi-objective optimization, there are often conflicts between different objective functions, making it challenging to find a solution vector $\underline{x}$ in the feasible space S that simultaneously optimizes all k objectives.

Optimization methods span a wide range of scientific and engineering fields, addressing diverse types and characteristics of problems. Broadly, optimization problems are divided into two categories: function optimization and combinatorial optimization. Function optimization typically involves continuous variables within a given domain, while combinatorial optimization deals with finding optimal solutions from a discrete set of possibilities (often represented as states or structures).

1.1.1   Local Optimization and Global Optimization

Definition 1.1 (Local Minimum) [1]: Let S is the search space defined by the constraint functions. If there exists a neighborhood B around x_B^*, for all $\forall x \in B$, it holds that:

$$f(x_B^*) \le f(x), \quad \forall x \in B \tag{1.4}$$

where $B \subset S \subseteq R^n$, then x_B^* is a local minimum point of $f(x)$ in B, and $f(x_B^*)$ is the corresponding local minimum value.

Most conventional optimization methods are designed for local optimization. Starting from a initial point $x_0 \in S$, these methods iteratively search for solutions that further improve the objective function until a specified stopping criterion is reached.

Definition 1.2 (Global Minimum) [1]: Let $S \in R^n$ is the search space defined by the constraints. If there exists $x_B^* \in S$, for all $\forall x \in S$, it holds that:

$$f(x^*) \leq f(x), \quad x \in S \tag{1.5}$$

where, then x^* is a global minimum point of $f(x)$ within S, and $f(x^*)$ is the global minimum value.

In convex programming, where both the objective function and feasible region are convex, any local optima is also a global optimium. However, for non-convex problems, where the objective function can have multiple local peaks within the feasible region, local optima may differ significantly from global optima.

While many algorithms for global optimization have been developed, there remains a considerable gap in maturity and robustness compared to the well-established methods for local optimization. To address global optimization, researchers are increasingly exploring stochastic optimization methods, such as simulated annealing, evolutionary algorithms, and swarm intelligence. These heuristic approaches offer effective and broadly applicable frameworks for tackling global optimization.

1.1.2 No Free Lunch Theorem

In optimization theory, a key contribution is the No Free Lunch (NFL) theorem proposed by Wolpert and Macready in 1997 [2]. It is supposed that both the search space X and the fitness landscape $Y \subseteq R$ are finite. Let $Y^X = \{F|F: X \to Y\}$ denote the set of all functions to be optimized. For any given X and Y, the average performance of any two optimization algorithms is identical when measured over all $f \in Y^X$.

This theorem implies that, when considering the entire set of possible functions, all optimization algorithms perform equally well. In other words, averaged across all possible problems, no single optimization algorithm is inherently superior to any other, including purely random search.

The key point behind the NFL theorem is the consideration of the "set of all possible functions," which includes deceptive functions and purely random functions. These functions often lack exploitable structure, exhibiting behavior that is largely random or misleading to typical search heuristics. Consequently, finding the optimal solution is like searching for a needle in a haystack, making it difficult for any specific search algorithm to consistently outperform others. Performance on such functions might even be worse than that of a pure random search.

However, there is also a third category. Functions in this category typically exhibit exploitable structures, patterns, or regularities in their value distribution that can provide guidance for locating the optimal point. Fortunately, most real-world optimization "problems" belong to this third category function.

Although the NFL theorem suggests that no optimization algorithm is universally superior across all possible functions, this conclusion does not necessarily hold when considering specific *subsets* or *classes* of functions. Christensen [3] proposed defining a classes of "optimizable" search functions, demonstrating general algorithms can provably outperform random search within this subset. In practice, the NFL theorem only negates the possibility of finding a single, universally optimal algorithm. However, for specific subset of functions (particularly those in the third category mentioned above), the theorem implies that it is indeed possible to develop algorithms that outperform random search.

1.2 Overview of Swarm Intelligence

After experiencing a decade of rapid progress in the 1980s, Artificial Intelligence (AI) encountered another downturn due, in part, to limitations inherent in its reliance on purely classical computational paradigms. However, as our understanding of the essence of life deepened, life sciences advanced at an unprecedented pace, enabling AI research to break free from the classical limitations of logical computation and explore new computational approaches. As AI pioneer *Marvin Minsky* suggested, "Biology provides a richer set of mechanisms for understanding intelligence than physics does." Consequently, research into biologically inspired computing ushered in a new era for AI. Against this backdrop, the self-organizing behaviors of social animals (e.g., ant colonies, fish schools, and bird flocks) attracted widespread attention. Many scholars modeled these behaviors mathematically and simulated on computers, leading to the emergence of what is now known as "swarm intelligence" [4].

The remarkable aspect of social animals is that while individual behaviors are simple, their collective actions can produce highly complex (and seemingly intelligent) behaviors. For example, a single ant has very limited abilities, but when simple ants form a colony, they can collectively accomplish complex tasks such as building nests, foraging, migrating, and cleaning the nest. Similarly, a seemingly disorganized swarm of bees can construct intricate hives, and flocks of birds can synchronize their flight without any centralized control [5].

Inspired by the mechanisms of biological evolution, the field of bionics was established in the mid-1950s and various new methods for solving complex problems were developed, such as genetic algorithms, evolutionary programming, and evolutionary strategies. As an innovative technology in evolutionary computation, swarm intelligence algorithms have increasingly attracted the attention of researchers. These algorithms have conceptual links with artificial life, particularly with evolutionary strategies and genetic algorithms. In swarm intelligence, a "swarm" refers to "a group of agents that can communicate directly or indirectly (e.g., by modifying the local environment) and work together to solve distributed problems" [5]. Swarm intelligence is thus defined as "the property of exhibiting intelligent behavior through the cooperation of non-intelligent agents" [5]. It provides a foundation for solving complex distributed problems without requiring centralized control or global models. Currently, representative swarm intelligence algorithms include Ant Colony Optimization (ACO) and Particle Swarm Optimization (PSO). ACO simulates the food-foraging behavior of ant colonies and has been successfully applied to discrete optimization. PSO inspired by social systems and bird flock foraging, has proven to be an effective optimization tool. Li et al. [6–8] have developed the Fish Swarm (FS) algorithm based on the characteristics of fish, which is also a type of swarm intelligence algorithm.

Swarm intelligence algorithms are oftern easy to implement, with its algorithms involve only basic mathematical operations and have relatively low requirements for CPU and memory. Moreover, these methods generally only require the output values of the objective function, not its gradient information, making them suitable for non-differentiable or "black-box" problems. Research into theories and applications of swarm intelligence has shown that these methods can effectively solve many global optimization problems. Importantly, the inherent parallelism and distributed nature of swarm intelligence provide a technological basis for handling large-scale problems and processing large volumes of data, potentially in distributed environments. From both theoretical and practical perspectives, the study and application of swarm intelligence hold significant academic and practical value.

1.2.1 Basic Principles and Characteristics of Swarm Intelligence

Rooted in the collective behavior of social animals and the theory of artificial life, swarm intelligence explores the fundamental principles underlying group behaviors and leverages these principles to design innovative problem-solving methods.

In 1994, Millonas [9] outlined five basic principles that swarm intelligence should ideally follow:

① **Proximity Principle**: The swarm should be capable of performing simple spatial and temporal computations. Since space and time can be linked to energy consumption, the swarm must be able to evaluate the utility of its response to a given spatiotemporal environment. This computation can be viewed as a direct behavioral response to environmental stimuli, which, to some extent, aims to maximizes the utility of certain collective behaviors.

② **Quality Principle**: The swarm should respond not only to spatial and temporal factors but also to quality factors in the environment, such as quality of resources or the safety of locations;

③ **Principle of Diverse Response**: The swarm should avoid narrowing its ways of resource acquisition. Instead, it should diversify its approaches to effectively prepare for sudden environmental changes. A perfectly ordered response to the environment, may be possible, but it is generally not desirable;

④ **Stability Principle**: The swarm should not change its behavior patterns with every environmental fluctuation. Frequent alterations consume energy and doesn't always yield valuable returns;

⑤ **Adaptability Principle**: The swarm should adapt its behavior patterns when the benefits outweigh the energy costs.

The stability and adaptability principles represent two facets of the same concept. The optimal response likely lies somewhere between complete order and total chaos. Therefore, a degree of randomness or diversity within the swarm's behavior is important: a moderate level can enhance adaptability and exploration, while excessive randomness may disrupt effecitive cooperation. It's important to note that these principles outline fundamental features often associated with swarm intelligence but are not necessarily exhaustive or definitive definitions. Generally, collective behaviors satisfying several of these principles can be considered indicative of swarm intelligence.

These principles suggest that, for intelligent agents to exhibit swarm intelligence, they must demonstrate autonomy, responsiveness, learning ability, and environment adaptability. The essence of swarm intelligence lies in the ability of a swarm of relatively simple agents to achieve complex group-level functions or complete a task through simple forms of interations and cooperation. "Simple agents" typically refers to individuals with limited capabilities or intelligence, while "simple cooperation" often involves local interations, such as direct communication with nearby agents or indirect communication via environmental modification (stigmergy), leading to mutual influence and coordinated actions. Swarm intelligence is characterized by the following attributes [10]:

① **Decentralized Control**: Control is decentralized, with no central controller. This typically makes the system more adaptable to dynamic environments and highly robust; for instance, the failure of one or a few individuals does not compromise the ability of the swarm to solve the overall problem;

② **Indirect Communication (Stigmergy)**: Each individual within the swarm can interact with the environment, enabling indirect communication between members—a process known as "*stigmergy*". Because swarm intelligence oftern relies on this form of indirect communication for information exchange and cooperation, the communication overhead typically increases minimal as the number of individuals grows. Thus, swarm intelligence possess excellent scalability;

③ **Simplicity of Agents**: The rules governing each individual's behavior, or their capabilities, are very generally very simple. Consequently, the implementation of swarm intelligence algorithms tends to be straightforward;

④ **Emergent Behavior**: The complex behaviors observed within the swarm arise ('emerge') from the local interactions of these simple agents, resulting in what is known as "Emergent Intelligence." This endows the system with self-organizing properties.

1.2.2 Ant Colony Optimization

Ant colony optimization (ACO) [4, 11] is a metaheuristic algorithm inspired by the foraging behavior of real ants, introduced in 1991 by Italian scholars *Alberto Colorni*, *Marco Dorigo*, and *Vittorio Maniezzo*. The algorithm is modeled on the way ant colonies find efficient paths between their nest and food sources using pheromone trails. ACO doesn't require any prior knowledge about the problem structure and begins by somewhat randomly exploring search paths. As the algorithm progresses and accumulates information (via simulated pheromones) about the solution space (e.g., paths), the search becomes more organized, gradually converging toward high-quality (often near-optimal or optimal) solutions. The learning mechanism of ACO involves three aspects:

① **Ant Memory**: Once an ant explores a path, it tends to choose the same path in future searches. This behavior is simulated in ACO with the use of a "Tabu" list;

② **Pheromone Communication**: Ants communicate with each other using pheromones. When an ant selects a path, it releases a chemical substance known as pheromones along the path. Other ants, when choosing their paths, are influenced by these pheromones.

③ **Collective Behavior**: It is challenging for a single ant to reach a food source, but the situation is quite different for the entire ant colony. As more ants traverse certain paths, the amount of pheromone left on those paths increases, which in turn enhances the pheromone intensity. Higher pheromone intensity raises the likelihood that other ants will choose that path, thus further boosting the pheromone levels. Conversely, on paths with fewer ants, the pheromones evaporate over time. By simulating this phenomenon, swarm intelligence can be used to establish a path selection mechanism, thereby guiding ACO towards the optimal solution.

The fundamental ACO model is expressed through the three equations below:

$$p_{ij}^k = \frac{\tau_{ij}^\alpha \cdot \eta_{ij}^\beta}{\sum\limits_{j \in A} \tau_{ij}^\alpha \cdot \eta_{ij}^\beta} \tag{1.6}$$

$$\tau_{ij}(n+1) = \rho \cdot \tau_{ij}(n) + \sum_{k=1}^{m} \Delta \tau_{ij}^k \tag{1.7}$$

$$\text{if } \Delta \tau_{ij}^k = \frac{Q}{\sum L_k}, \quad \text{the } k \text{ - th ant traverses the path from } i \text{ to } j \tag{1.8}$$

The parameters in Eqs. (1.6), (1.7), and (1.8) are defined as follows: m is the number of ants; n is the number of iterations; i is the current position of the ant; j is the possible position that can be reached; A is the set of possible positions for the ant; p_{ij}^k is the probability of the k-th ant moving from position i to position j; η_{ij} is the heuristic information, representing the visibility of the path from i to j, commonly expressed as $\eta_{ij} = 1/d_{ij}$, where d_{ij} represents the distance between positions; α and β are weights for pheromone and heuristic information, respectively; τ_{ij} is the pheromone intensity on the path from i to j; $\Delta\tau_{ij}^k$ is the amount of pheromone deposited per unit length by the k-th ant on the path from i to j; ρ is the evaporation coefficient of pheromone; Q is the quality coefficient of pheromone; L_k is the objective function (e.g., the *Euclidean* distance between positions).

Currently, computational intelligence methodologies, including ant colony optimization, genetic algorithms, simulated annealing, and tabu search, have demonstrated exceptional efficacy in a variety of combinatorial optimization problems. These includes graph coloring, flow-shop scheduling, vehicle routing, robotic path planning, and routing algorithm design. Additionally, ACO has also exhibited significant performance in domains such as function optimization, system identification, and data mining. As a seminal paradigm in swarm intelligence research, ACO was once widely regarded as a defining framework for the field. In recent years, continuous theoretical advancements in swarm intelligence principles and algorithmic refinements have propelled the emergence of novel breakthroughs, supporting the field's momentum for innovation.

1.2.3 Particle Swarm Optimization

Particle swarm optimization (PSO), proposed by Dr. Kennedy and Prof. Eberhart in 1995[12], is a stochastic, parallel optimization algorithm. It is simple in concept, easy to implement, and does not require the objective function to be differentiable, continuous, or other specific properties. Additionally, PSO has a fast convergence rate. Since then, PSO has been extensively developed and successfully applied in various fields, including single-objective optimization, constrained optimization, dynamic optimization, multi-objective optimization, and dynamic multi-objective optimization.

The basic concept of PSO is inspired by the foraging behavior of bird flocks. Consider a scenario where a flock of birds searches randomly for food within a specific area, unaware of the food's exact location or the distance from their current position. The birds' strategy, in this case, is to follow the lead bird in the flock and search collectively. PSO draws inspiration from this model that each candidate solution to an optimization problem is treated as a bird in the search space, referred to as a "particle", moreover each particle has a fitness value, determined by the objective function. These particles move through the solution space with a random velocity. By exchanging information with others, they acquire heuristic guidance from one another, which collectively directs the movement of the entire swarm.

1.3 The Current State and Applications of the PSO Algorithm

Similar to genetic algorithms (GAs), PSO is an iterative optimization technique, but it does not employ the *crossover* and *mutation* operators used in genetic algorithms. The swarm in PSO is initialized as a group of random particles (random solutions) that search for the optimal value through iterative updates. In each iteration, particles adjust their positions by tracking two "extrema":

- *pbest* (personal best particle), the best solution found by each particle individually,
- *gbest* (global best particle), the best solution found by the entire swarm.

Like GAs, PSO has made significant progress since its introduction. According to a statistical analysis by Riccardo [13], based on papers retrieved from IEEE Xplore, as of 2007, approximately two-thirds of over 1,000 PSO-related research papers focused on its applications, with Particle Swarm Optimization (PSO) demonstrating strong performance in optimization computing, biomedical fields, communication networks, control systems, and pattern classification. The remaining one-third of the papers explored enhancements and analyses of PSO, particularly in areas such as parameter selection, stability, convergence, and statistical characteristics. Figure 1.1a presents the number of PSO-related papers published in *IEEE Xplore* during the the first decade starting from 1995, while Fig. 1.1b illustrates the publication trends over the past 20 years. These figures demonstrate that researchers continue to exhibit strong enthusiasm for the study and application of Particle Swarm Optimization (PSO).

1.3.1 Theoretical Analysis of PSO

The process by which PSO searches for the optimal solution within the feasible solution space is inherently stochastic. Consequently, research based on stochastic process theory is one of the primary approaches to studying the convergence of PSO.

From the perspective of the stochastic process, Jin et al. [14] transformed the dynamic PSO model with stochastic factors into a linear time-invariant system in a probabilistic sense and provided a sufficient condition for the probability of convergence.

Yuan and Chen [15] developed a mathematical model of the algorithm's Markov chain process.

Li [16] proved that the chain of particle states exhibits a Markov chain property and that the state sequence of the swarm constitutes a finite homogeneous Markov chain. Li also analyzed particle trajectories from the perspective of difference equations, though the study did not further elaborate on using Markov chains to study PSO convergence.

Fig. 1.1 Statistics of
PSO-related papers

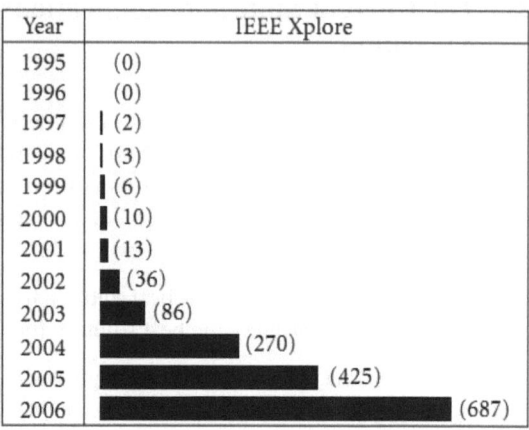

(a) papers in the first decade starting from 1995

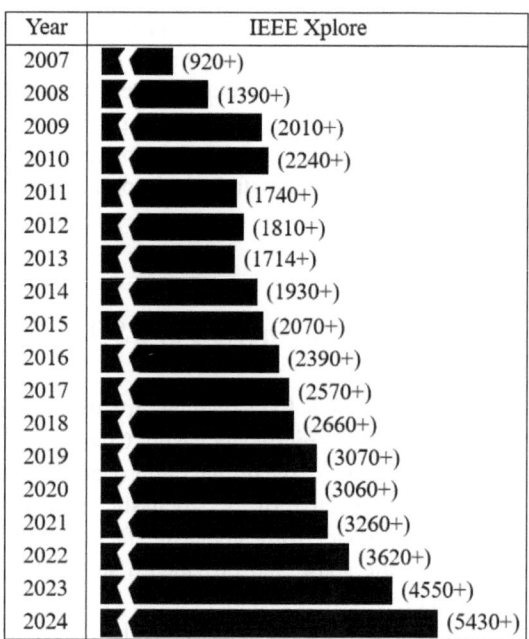

(b) papers in the recent 20 years

Ren et al. [17] approached the problem by examining the Markov chain formed by particle states, demonstrating the reducibility and non-homogeneity of the particle state space, and verifying that the particle state space is transient, thereby confirming that the Markov chain in PSO lacks a steady state, implying the absence of global convergence.

Cai [18] established an absorbing Markov process for PSO and introduced the reachable state set as a key indicator for convergence analysis, moreover proposed methods to improve the global convergence of PSO. However, a complete proof of convergence analysis for the standard PSO remains unestablished.

Poli [19, 20] constructed a discrete Markov chain for Bare Bones Particle Swarm (BBPS) using finite element grids and demonstrated that this model can approximate continuous problems with arbitrary precision. The transition matrix of the iterative chain provides precise information on particle behavior in each generation of BBPS, including the probability of finding the global optimum or reaching an erroneous solution. Poli also explained the characteristics of applying Cauchy, Gaussian, and other probability distributions for the acceleration coefficients in the standard PSO.

Pan et al. [21] defined particle and swarm states based on the difference model of PSO and analyzed their Markov properties, proving the closedness of the optimal state sets for particles and swarms, and computing the one-step transition probabilities for particles. Further, using the total probability formula and Markov chain, they derived the transition probabilities from swarm states to the optimal state set. Based on these transition probabilities, they discussed the roles of the inertia weight ω and the acceleration coefficient c of PSO, investigated issues like premature convergence and divergence, and ultimately demonstrated that standard PSO converges to the *gbest* with a certain probability.

Research on the parameter selection and stability of PSO primarily is grounded in various stability theories of dynamic systems.

Trelea [22] utilized the theory of linear dynamic systems with constant coefficients to examine the conditions for parameter selection and stability in PSO.

Clerc [23] developed a constrained PSO defined by five parameters and analyzed its convergence properties and particle trajectory characteristics in the phase plane.

Zeng et al. [24] proposed a continuous PSO and provided conditions for the asymptotic convergence of the evolution equation under different parameter settings.

Cui et al. [25] further introduced an enhanced PSO algorithm corresponding to a system composed of integral and oscillatory components, discussing methods for parameter selection. However, the dynamic equation governing particle motion in PSO is not a linear time-invariant system, which limits the applicability of analyses based on time-invariant dynamic systems.

Kadirkamanathan [26] characterized the *gbest* particle, which entered a stagnation phase, as a nonlinear feedback system and analyzed the sufficient conditions for PSO stability based on the Lyapunov stability theorem. Due to the conservative nature of this analysis, the parameter selection becomes very stringent, allowing particles to maintain minimal and local oscillations only.

Martínez [27, 28] demonstrated that PSO can be physically represented as a discrete, stochastic, damped mass-spring system and inferred that different PSO models perform differently in balancing exploration and exploitation, with their convergence related to the stability of first- and second-order systems.

Pan et al. [29] treated PSO as a dynamic time-varying system, analyzing the sufficient conditions for its stability without the constraint of Lipschitz conditions and extending the range of the inertia weight ω to $(-1, 1)$. They further examined the algorithm's search performance when the swarm stagnates at a local extremum in the solution space [29–32] and its similarity to BBPS [33].

From the perspective of fundamental stochastic optimization frameworks, Bergh [34, 35] demonstrated that although the PSO algorithm exhibits excellent performance it is not a global optimization method and cannot even guarantee convergence to a local optimum. He also proposed methods to ensure the local convergence.

Jiang et al. [36] examined how different parameters influence the stochastic convergence of the algorithm.

Poli [37] investigated the statistical characteristics, including the expected value and variance of swarm positions, when the *gbest* solution stagnated, analyzing how parameter settings impact these statistical traits.

Zhang et al. [38] provided a detailed explanation of the global convergence criteria for stochastic optimization and applied these criteria to theoretically analyze PSO's the global convergence, highlighting that PSO does not satisfy the global convergence criteria for stochastic optimization algorithms. Zhang [39] analyzed how the eigenvalues distribution in the PSO system, based on discrete system theory, affects the convergence trajectory and proposed parameter selection principles from the perspective of dynamic system characteristics.

1.3.2   Enhancement Strategies for PSO

To address the issues of premature convergence and to increase swarm diversity for improved optimization performance, researchers have developed various enhancement strategies.

Since the settings of inertia weight and acceleration coefficients significantly influence algorithm effectiveness, Numerous enhancement strategies have been proposed. For the inertia weight, Shi and Eberhart introduced linear decreasing inertia [40], random adjustment [41], and fuzzy adaptive strategies [42]. Nickabadi et al. [43] suggested adjusting the inertia weight based on the proportion of successful particles in the swarm, while Zhou et al. [44] proposed tuning it according to particle velocity. Saha et al. [45] developed an inertia weight adjustment strategy considering both the position and velocity of particles. Ting et al. [46] introduced an exponential adjustment method of the inertia weight to enhance algorithm performance, and Yang et al. [47] proposed a higher-order nonlinear time-varying inertia adjustment. Taherkhani et al. [48], based on swarm stability, developed an adaptive inertia weight adjustment strategy.

Regarding acceleration coefficients, Clerc [49] presented an adaptive adjustment method using the fitness distance between individual particle and the swarm as a basis. Juang et al. [50] proposed an adaptive fuzzy adjustment to improve how PSO solves multi-modal functions. Ali [51] suggested an exponentially iterated adjustment for the acceleration coefficient, while Ismail and Engelbrecht [52] developed an adaptive learning strategy for parameters tuning. Abedinia et al. [53] introduced a PSO variant with time-varying acceleration coefficients to solve economic load dispatch problems. Chen et al. [54] presented a sine–cosine acceleration coefficient to balance exploration and convergence, and Liu et al. [55] proposed an adaptive acceleration coefficient adjustment using a weighted sigmoid function.

To prevent premature convergence, Kennedy [56] and Das [57] first proposed the use of neighborhood topology. Subsequently, Kennedy and Mendes [58] investigated the impact of various topological structures on algorithm performance. *Mendes* [59], in further studies, discussed the effects of different static topologies on algorithm performance and introduced Fully Informed Particle Swarm (FIPS) model. Janson and Middendorf [60] proposed a dynamic hierarchical topology. Liang and Suganthan [61] introduced a multi-swarm strategy effectively functioning as a dynamic neighborhood topology. Qu et al. [62] developed a niche-based dynamic neighborhood topology. Ghosh et al. [63] designed a hybrid PSO with varialble multi-sub-swarm topologies, where the number of sub-swarms is variable, and the number of particles within each sub-swarm changes dynamically. Wang et al. [64] introduced a dynamic tournament topology, allowing non-gbest particles to probabilistically become leader and, enhance intra-swarm information exchange.

In addition to improve PSO through parameter adjustment or changes in topology, researchers have also developed various hybrid PSO by incorporating ideas from other optimization algorithms. The combination of PSO with genetic algorithms has been successfully applied to solve complex problems such as dynamic clustering [65], automated generation of software testing data [66], reservoir permeability prediction in neural network training [67], multi-satellite observation scheduling [68], closed-loop supply chain planning [69], and photovoltaic power generation cost optimization [70] PSO combined with simulated annealing has addressed optimization tasks including cluster analysis [71], urban water demand prediction [72], SVM optimization [73], dynamic spectrum allocation in broadcast networks [74], dynamic optimal power allocation in power systems [75], and train timetable optimization [76]. Xin et al. [77] conducted a comprehensive review of differential evolution (DE)-PSO hybrid algorithms. Concluding that their effective integration balances exploration and exploitation capabilities. The combination of PSO with ACO has been used for multi-attribute partner selection in virtual enterprises [78], Turkey's energy demand forecasting [79], multi-processor task scheduling [80], and planar truss structure optimization [81].

1.3.3 Current Status of PSO Applications

Since its introduction in 1995, PSO has attracted significant attention due to its simplicity, clear principles, and practical advantages. Consequently, substantial progress has been made in both theoretical research and real-world applications across diverse fields.

In engineering, PSO has been extensively used. Robinson [82] applied it to antenna design in communication engineering; Li [83] used it for vehicle path planning; Bergh [84] employed collaborative PSO for training neural networks in pattern classification; Hou [85] designed IIR digital low-pass and band-pass filters using PSO; Long [86] applied PSO to generator parameters identification; Xia et al. [87] used PSO for the optimal design of linear induction motors; PSO has also been applied to optimize intelligent controllers in automatic control systems, system identification, and road spectrum modeling [88]. He [89] optimized fuzzy control systems and designed fuzzy controllers by PSO. Delgarm et al. [90] applied both single-objective and multi-objective PSO to building energy simulation. Nouiri et al. [91] used PSO to solve flexible job shop scheduling problems. Additionally, PSO has broad applications in the of power systems optimization [92–96], including expansion and planning of distribution networks, maintenance scheduling, unit commitment, load economic dispatch, optimal power flow and reactive power control, harmonic analysis and capacitor placement, distribution network state estimation, parameter identification, and optimal design.

PSO has also been widely adopted in other domains. Gao et al. [97] adapted and modified PSO to solve the traveling salesman problem (TSP). Zhang et al. [98] used binary PSO with mutation operators for email classification. Ting [99] applied a hybrid PSO, combining binary- and real-coded encodings to solve unit commitment optimizations. With binary-coded PSO for unit commitment optimization and real-coded PSO, they optimized economic load distributions. Other applications include e-commerce pricing [100], constrained layout optimization [101], portfolio investment of new products [102], and advertising resource optimization [103].

1.4 Summary

This chapter provides an overview of swarm intelligence algorithms and defines optimization problems. It starts with a general discussion of computational intelligence, followed by an explanation of optimization problem concepts. A brief introduction to swarm intelligence algorithms is provided, culminating in a detailed analysis of the theoretical foundations, enhancement strategies, and application status of PSO.

References

1. Zeng JH, Jie J, Cui ZH (2004) Particle swarm optimization [M]. Science Press, Beijing
2. Wolpert DH, Macready WG (1997) No free lunch theorems for optimization [J]. IEEE Trans Evol Comput 1(1):67–82

3. Christensen S, Oppacher F (2001) What can we learn from no free lunch? A first attempt to characterize the concept of searchable function. In: Proceedings of the the 2001 conference on genetic and computation
4. Gao S, Yang JY (2006) Swarm intelligence algorithms and applications [M]. China Water & Power Press, Beijing
5. Chen ZY (2009) Research on neighbor topology structure and algorithm improvement of particle swarm optimization. Chongqing University
6. Li XL (2003) An optimizing method based on autonomous animats: fish-swarm algorithm. Zhejiang University
7. Li XL, Shao ZJ, Qian JX (2002) An optimizing method based on autonomous animats: fish-swarm algorithm. Syst Eng-Theory Pract 11:32–38
8. Li XL, Qian JX (2001) Artificial fish swarm algorithm: a bottom-up optimization model. In: Proceedings of the 2001 conference on process systems engineering
9. Millonas MM (1992) United States
10. Zhang Y, Kang Q, Wang L, et al Swarm intelligence [J/OL]
11. Dorigo M, Birattari M, Stutzle T (2007) Ant colony optimization. Tsinghua University Press, Beijing
12. Kennedy J, Eberhart R (1995) Particle swarm optimization. In: Proceedings of the 1995 IEEE International conference on neural networks part 1 (of 6), November 27, 1995–December 1, 1995. IEEE, Perth, Aust
13. Poli R (2008) Analysis of the publications on the applications of particle swarm optimisation. J Artif Evolut Appl 2008(1):685175
14. Jin XL, Ma LH, Wu TJ, et al (2007) Convergence analysis of the particle swarm optimization based on stochastic processes [J]. Acta Automat Sin 12:1263–1268
15. Yuan DL, Chen Q (2009) Particle swarm optimization algorithm based on markov model and its stochastic process analysis. Comput Eng Appl 31:49–52
16. Li N (2006) Theoretical analysis and application research of particle swarm optimization algorithm. Huazhong University of Science and Technology
17. Ren ZH, Wang J, Gao YL (2011) The global convergence analysis of particle swarm optimization algorithm based on Markov chain. Control Theory Appl 4:462–466
18. Cai ZQ, Huang H, Zheng ZH, et al (2009) Convergence improvement of particle swarm optimization based on the expanding attaining-state set. J Huazhong Univ Sci Technol (Nat Sci Edn) 6:44–47
19. Riccardo PWBL (2007) Markov chain models of bare-bones particle swarm optimizers [J]. In: Proceedings of the 9th annual conference on genetic and evolutionary computation. ACM, London, England, pp 142–149
20. Poli R, Langdon WB (2007) Markov chain models of bare-bones particle swarm optimizers. In: Proceedings of the 9th annual genetic and evolutionary computation conference, GECCO 2007, July 7, 2007–July 11, 2007. Association for Computing Machinery, London, United kingdom
21. Pan F, Zhou Q, Li WX, et al Analysis of standard particle swarm optimization algorithm based on Markov chain [J]. Acta Automat Sin
22. Trelea IC (2003) The particle swarm optimization algorithm: convergence analysis and parameter selection [J]. Inf Process Lett 85(6):317–325
23. Clerc M, Kennedy J (2002) The particle swarm—explosion, stability, and convergence in a multidimensional complex space [J]. IEEE Trans Evol Comput 6(1):58–73
24. Zeng JC, Cui ZH (2006) A new unified model of particle swarm optimization and its theoretical analysis. J Comput Res Dev 1:96–100
25. Cui ZH, Zeng JC (2006) Analysis and improvement about particle swarm optimization based on linear control theory. J Chinese Comput Syst 5:849–853
26. Kadirkamanathan V, Selvarajah K, Fleming PJ (2006) Stability analysis of the particle dynamics in particle swarm optimizer [J]. IEEE Trans Evol Comput 10(3):245–255
27. Fernández-Martínez JL, García-Gonzalo E, Saraswathi S, Jernigan R, Kloczkowski A (2011) Particle swarm optimization: a powerful family of stochastic optimizers. Analysis, design and application to inverse modelling. pp 1–8

28. Fernandez-Martinez JL, Garcia-Gonzalo E, Saraswathi S, et al (2011) Particle swarm optimization: a powerful family of stochastic optimizers. Analysis, design and application to inverse modelling. In: Proceedings of the 2nd International conference on swarm intelligence, ICSI 2011, June 12, 2011–June 15, 2011. Springer Verlag, Chongqing, China
29. Pan F, Chen J, Gan GM, et al (2006) Model analysis of particle swarm optimizer [J]. Acta Automat Sin 3:368–377
30. Clerc M (2006) Stagnation analysis in particle swarm optimization or what happens when nothing happens http://clerc.maurice.free.fr/pso/
31. Ming J, Yupin L, Shiyuan Y (2007) Stagnation analysis in particle swarm optimization. In: Proceedings of the 2007 IEEE swarm intelligence symposium. IEEE, Piscataway, NJ, USA
32. Poli R, Broomhead D (2007) Exact analysis of the sampling distribution for the canonical particle swarm optimiser and its convergence during stagnation. In: Proceedings of the 9th annual genetic and evolutionary computation conference, GECCO 2007, July 7, 2007–July 11, 2007. Association for Computing Machinery, London, United kingdom
33. Pan F, Chen J, Xin B, et al (2009) Several characteristics analysis of particle swarm optimizer [J]. Acta Automat Sin 7:1010–1016
34. van den Bergh F (2002) An analysis of particle swarm optimizers [D]. University of Pretoria, South Africa
35. van den Bergh F, Engelbrecht AP (2006) A study of particle swarm optimization particle trajectories [J]. Inf Sci 176(8):937–971
36. Jiang M, Luo YP, Yang SY (2007) Stochastic convergence analysis and parameter selection of the standard particle swarm optimization algorithm [J]. Inf Process Lett 102(1):8–16
37. Poli R (2009) Mean and variance of the sampling distribution of particle swarm optimizers during stagnation [J]. IEEE Trans Evol Comput 13(4):712–721
38. Zhang HB, Wang HB, Hu ZJ (2011) Analysis of particle swarm optimization algorithm global convergence method [J]. Comput Eng Appl 34:61–63
39. Zhang W (2010) Research on particle swarm optimization algorithm and its application in array antennas. Taiyuan University of Technology
40. Shi Y, Eberhart RC (1999) Empirical study of particle swarm optimization. In: Proceedings of the 1999 congress on evolutionary computation-CEC99, 6–9 July 1999. IEEE, Piscataway, NJ, USA
41. Eberhart RC, Yuhui S (2001) Tracking and optimizing dynamic systems with particle swarms. In: Proceedings of the 2001 congress on evolutionary computation. IEEE, Piscataway, NJ, USA
42. Yuhui S, Eberhart RC (2001) Fuzzy adaptive particle swarm optimization. In: Proceedings of the 2001 congress on evolutionary computation. IEEE, Piscataway, NJ, USA
43. Nickabadi A, Ebadzadeh MM, Safabakhsh R (2011) A novel particle swarm optimization algorithm with adaptive inertia weight [J]. Appl Soft Comput 11(4):3658–3670
44. Zheng Z, Yuhui S (2011) Inertia weight adaption in particle swarm optimization algorithm. In: Proceedings of the advances in swarm intelligence second international conference, ICSI 2011. Springer Verlag, Berlin, Germany
45. Saha SK, Sarkar S, Kar R, et al (2012) Digital stable IIR low pass filter optimization using particle swarm optimization with improved inertia weight. In: Proceedings of the 2012 international joint conference on computer science and software engineering (JCSSE 2012), 30 May–1 June 2012. IEEE Computer Society, Los Alamitos, CA, USA
46. Ting TO, Yuhui S, Shi C, et al (2012) Exponential inertia weight for particle swarm optimization. In: Proceedings of the advances in swarm intelligence third international conference, ICSI 2012, 17–20 June 2012. Springer-Verlag, Berlin, Germany
47. Yang CW, Gao W, Liu NG et al (2015) Low-discrepancy sequence initialized particle swarm optimization algorithm with high-order nonlinear time-varying inertia weight[J]. Appl Soft Comput 29:386–394
48. Taherkhani M, Safabakhsh R (2016) A novel stability-based adaptive inertia weight for particle swarm optimization[J]. Appl Soft Comput 38:281–295

49. Locally MCT (2001) Act locally: the way of life of cheap-pso, an adaptive PSO. http://clerc. maurice.free.fr/pso
50. Yau-Tarng J, Shen-Lung T, Hung-Chih C (2011) Adaptive fuzzy particle swarm optimization for global optimization of multimodal functions [J]. Inf Sci 181(20):4539–4549
51. Ben Ali YM (2011) An augmented particle swarm model based bi-acceleration factor [J]. Int J Intell Comput Cybernet 4(2):187–205
52. Ismail A, Engelbrecht AP (2012) The self-adaptive comprehensive learning particle swarm optimizer. In: Proceedings of the swarm intelligence 8th international conference, ANTS 2012, 12–14 Sept 2012. Springer Verlag, Berlin, Germany
53. Abedinia O, Amjady N, Ghasemi A, et al (2012) Solution of economic load dispatch problem via hybrid particle swarm optimization with time-varying acceleration coefficients and bacteria foraging algorithm techniques
54. Chen K, Zhou FY, Yin L et al (2018) A hybrid particle swarm optimizer with sine cosine acceleration coefficients[J]. Inf Sci 422:218–241
55. Liu WB, Wang ZD, Yuan Y et al (2021) A novel sigmoid-function-based adaptive weighted particle swarm optimizer[J]. IEEE Trans Cybernet 51(2):1085–1093
56. Kennedy J (1999) Small worlds and mega-minds: effects of neighborhood topology on particle swarm performance. In: Proceedings of the 1999 congress on evolutionary computation-CEC99. IEEE, Piscataway, NJ, USA
57. Suganthan PN (1999) Particle swarm optimiser with neighbourhood operator. In: Proceedings of the 1999 congress on evolutionary computation-CEC99, 6–9 July 1999. IEEE, Piscataway, NJ, USA
58. Kennedy J, Mendes R (2002) Population structure and particle swarm performance. In: Proceedings of the 2002 world congress on computational intelligence—WCCI'02. IEEE, Piscataway, NJ, USA
59. Mendes R (2004) Population topologies and their influence in particle swarm performance [D]. Publica: Universidade do Minho
60. Janson S, Middendorf M (2005) A hierarchical particle swarm optimizer and its adaptive variant [J]. IEEE Trans Syst Man Cybernet Part B (Cybernetics) 35(6):1272–1282
61. Liang JJ, Suganthan PN (2005) Dynamic multi-swarm particle swarm optimizer with local search. In: Proceedings of the 2005 IEEE congress on evolutionary computation, IEEE CEC 2005, September 2, 2005–September 5, 2005, Edinburgh, Scotland, United kingdom. Institute of Electrical and Electronics Engineers Computer Society.
62. Qu BY, Liang JJ, Suganthan PN (2012) Niching particle swarm optimization with local search for multi-modal optimization [J]. Inform Sci 197:131–143
63. Ghosh A, Chowdhury A, Sinha S, et al (2012) A genetic Lbest particle swarm optimizer with dynamically varying subswarm topology. In: Proceedings of the 2012 IEEE congress on evolutionary computation (CEC). IEEE, Piscataway, NJ, USA
64. Wang L, Yang B, Orchard J (2016) Particle swarm optimization using dynamic tournament topology[J]. Appl Soft Comput 48:584–596
65. Kuo RJ, Syu YJ, Chen Z-Y, et al (2012) Integration of particle swarm optimization and genetic algorithm for dynamic clustering [J]. Inform Sci 195:124–140
66. Rui D, Xianbin F, Shuping L, et al (2012) Automatic generation of software test data based on hybrid particle swarm genetic algorithm. In: Proceedings of the 2012 IEEE symposium on electrical and electronics engineering (EEESYM 2012). IEEE, Piscataway, NJ, USA
67. Ali AM, Zendehboudi S, Lohi A, et al (2012) Reservoir permeability prediction by neural networks combined with hybrid genetic algorithm and particle swarm optimization [M]. https://doi.org/10.1111/j.1365-2478.2012.01080.x
68. Chen Y, Zhang D, Zhou M, et al (2012) Multi-satellite observation scheduling algorithm based on hybrid genetic particle swarm optimization. In: Proceedings of the 2nd International conference of electrical and electronics engineering, ICEEE 2011, December 1, 2011–December 2, 2011. Springer Verlag, Macau, China
69. Soleimani H, Kannan G (2015) A hybrid particle swarm optimization and genetic algorithm for closed-loop supply chain network design in large-scale networks[J]. Appl Math Model 39(14):3990–4012

70. Ghorbani N, Kasaeian A, Toopshekan A et al (2018) Optimizing a hybrid wind-PV-battery system using GA-PSO and MOPSO for reducing cost and increasing reliability[J]. Energy 154:581–591

71. Zhang Y, Wu L (2012) Restarted simulated annealing particle swarm optimization used in cluster analysis [J]. Int J Res Rev Soft Intell Comput 2(3):201–206

72. Bo S, Jiancang X, Ni W (2012) Application of urban water demand prediction model by using particle swarm algorithm based on simulated annealing [J]. Appl Mech Mater 155–156:102–106

73. Jiao B, Xu ZX (2012) Nonlinear inertia weigh particle swarm optimization combines simulated annealing algorithm and application in function and SVM optimization [J]. Appl Mech Mater 130(1):3467–3471

74. Tang M, Long C, Guan X et al (2012) Nonconvex dynamic spectrum allocation for cognitive radio networks via particle swarm optimization and simulated annealing [J]. Comput Netw 56(11):2690–2699

75. Niknam T, Narimani MR, Jabbari M (2012) Dynamic optimal power flow using hybrid particle swarm optimization and simulated annealing [M]. Europ Trans Electric Power https://doi.org/10.1002/etep.1633

76. Guo X, Sun HJ, Wu JJ et al (2017) Multiperiod-based timetable optimization for metro transit networks[J]. Transport Res Part B Method 96:46–67

77. Xin B, Chen J, Zhang J, Fang H, Peng ZH (2012) Hybridizing differential evolution and particle swarm optimization to design powerful optimizers: a review and taxonomy [J]. IEEE Trans Syst Man Cybernet Part C Appl Rev 42(5):744–767

78. Niu SH, Ong SK, Nee AY (2011) A hybrid particle swarm and ant colony optimizer for multi-attribute partnership selection in virtual enterprises [M]. JohnWiley & Sons, Inc., pp 289–326

79. Kiran MS, Ozceylan E, Gunduz M et al (2012) A novel hybrid approach based on particle swarm optimization and ant colony algorithm to forecast energy demand of Turkey [J]. Energy Convers Manage 53(1):75–83

80. Thanushkodi K, Deeba K (2012) Hybrid intelligent algorithm [improved particle swarm optimization (PSO) with ant colony optimization (ACO)] for multiprocessor job scheduling [J]. Sci Res Essays 7(20):1935–1953

81. Gholizadeh S, Fattahi F (2012) Serial integration of particle swarm and ant colony algorithms for structural optimization [J]. Asian J Civil Eng (Build Housing) 13(1):127–146

82. Robinson J, Rahmat-Samii Y (2004) Particle swarm optimization in electromagnetics [J]. IEEE Trans Anten Propag 52(2):397–407

83. Li N, Zou T, Sun DB (2004) Particle swarm algorithm for vehicle routing problems with time windows [J]. Syst Eng-Theory Pract 4:130–135

84. Van den Bergh F, Engelbrecht AP (2001) Training product unit networks using cooperative particle swarm optimisers. In: Proceedings of the International joint conference on neural networks (IJCNN'01), July 15, 2001–July 19, 2001. Institute of Electrical and Electronics Engineers Inc., Washington, DC, United states

85. Hou ZR (2003) Improved particle swarm optimization algorithm and its application research in digital filter design. Master's Thesis, Lanzhou University, 3

86. Long Y, Wang JQ (2003) Parameters identification of synchronous generator based on particle swarm optimization theory [J]. Large Electr Mach Hydraul Turb 1:8–11

87. Xia YM, Fu ZY, Yuan SY, et al (2002) Application of particle swarm optimism in optimistic design for linear induction motor [J]. S&M Electr Mach 6:14–16

88. Pan F (2005) Coordinated research on theory and methods of particle swarm optimization and its application in servo systems. Doctoral Dissertation, Beijing Institute of Technology 8

89. He Z, Wei C, Yang L, et al (1998) Extracting rules from fuzzy neural network by particle swarm optimisation. In: Proceedings of the 1998 IEEE International conference on evolutionary computation, 1998 IEEE world congress on computational intelligence

90. Delgarm N, Sajadi B, Kowsary F (2016) Multi-objective optimization of the building energy performance: a simulation-based approach by means of particle swarm optimization (PSO). Appl Energy 170:293–303

91. Nouiri M, Bekrar A, Jemai A et al (2018) An effective and distributed particle swarm optimization algorithm for flexible job-shop scheduling problem. J Intell Manuf 29(3):603–615
92. Yoshida HKK, Fukuyama Y, et al (1999) A particle swarm optimization for reactive power and voltage control considering voltage stability [J]. In: Proceeding of the international conference oil intelligent system application to power systems. pp 117–121
93. Yu XM, Li Y, Xiong XG, Wu YW (2003) Optimal shunt capacitor placement using particle swarm optimization algorithm with harmonic distortion consideration [J]. Proc CSEE 23(2):26–31
94. Liu YZH, et al (2001) Particle swarm optimization for fault state power supply reliability enhancement [J]. In: Proceeding of IEEE international conference oil intelligent systems application to power systems, Budapest, Hungary. pp 172–176
95. Yoshida H, Kawata K, Fukuyama Y, et al (2001) A particle swarm optimization for reactive power and voltage control considering voltage security assessment. In: Proceedings of the power engineering society winter meeting. IEEE
96. Yu XM, Li Y, Xiong XG, et al (2003) Optimal shunt capacitor placement using particle swarm optimization algorithm with harmonic distortion consideration [J]. Proc CSEE 2:30–4+124
97. Gao S, Han B, Wu XJ, et al (2004) Solving traveling salesman problem by hybrid particle swarm optimization algorithm [J]. Control Decis 11:1286–1289
98. Zhan ZH, Li JJ, Cao JN (2013) Multiple populations for multiple objectives: a coevolutionary technique for solving multiobjective optimization problems[J]. IEEE Trans Cybernet 43(2):445–463
99. Ting TO, Rao MVC, Loo CK (2006) A novel approach for unit commitment problem via an effective hybrid particle swarm optimization [J]. IEEE Trans Power Syst 21(1):411–418
100. Wang JW (2006) Improvements and applications of particle swarm optimization algorithm [D]. Northeastern University
101. Li N, Liu F, Sun DB (2004) A study on the particle swarm optimization with mutation operator constrained layout optimization. Chinese J Comput 7:897–903
102. Wan FC, Wang DW, Li YP (2004) Particls swarm optimization of correlative product combinatorial introduction model. Control Decis 5:520–524
103. Qi J, Wang DW (2004) Particle swarm optimization algorithm for a model of optimally scheduling web advertising resources. Control Decis 8:881–884
104. Ji Z, Liao HL, Wu QH (2009) Particle swarm algorithm and applications [M]. Science Press, Beijing

Chapter 2
Overview of PSO

An optimization algorithm is a search process that applies optimization principles and mechanisms through specific methods and rules to find solutions to problems. In engineering, optimization methods primarily include classical algorithms, constructive algorithms, neighborhood search algorithms, system dynamics-based algorithms, and hybrid algorithms [1]. Proposed by Kennedy and Eberhart in 1995 [2, 3], Particle Swarm Optimization (PSO) is an evolutionary computation algorithm rooted in swarm intelligence. Like other evolutionary algorithms such as genetic algorithms (GA), evolutionary programming (EP), and evolution strategies (ES), PSO is a swarm-based direct search that does not rely on gradient or curvature information, but instead uses nature-inspired heuristics for optimization.

2.1 Fundamental Framework of Stochastic Search Algorithms

Back and Schwefel [4] proposed a unified framework for evolutionary computation [4–6], which is classified under stochastic search algorithms. These algorithms require a population of sample solutions, apply random variations to these solutions according to certain rules, and produce the next generation of the population through selection. In general, random search can be described as follows.

Definition 2.1 The objective function: $f: R^n \rightarrow R$, $S^n \subseteq R^n$, where the goal is to find a vector $x^* \in S^n$ that minimizes or maximizes $f(x)$ or to generate an acceptable solution within the feasible set S^n.

Based on Definition 2.1, the fundamental framework of a stochastic search algorithm can be formalized as follows.

Definition 2.2 - Stochastic Search Framework:

© Beijing Institute of Technology Press 2025
F. Pan et al., *Particle Swarm Optimizer and Multi-Objective Optimization*,
https://doi.org/10.1007/978-981-95-3381-7_2

(1) **Initialization**: Generate an initial solution $x^0 \in S^n$, initialize the optimization process, and set the iteration counter $k = 0$;

(2) **Sampling**: In the sampling space (R^n, B, μ_k), generate a new sample solution $\zeta^k \in S^n$ using a neighborhood function ϕ;

(3) **Selection**: Select a new probability measure μ_{k+1}, update the current solution as $x^{k+1} = D(x^0, \zeta^k)$, increment k, and return to (2).

Here, the neighborhood function $\phi: R^n \rightarrow R$ is a mapping that generates new sample solutions from the current solution; the selection operator $D: S^n \times R^n \rightarrow S^n$ determines the update rule; B denotes the Borel σ-algebra over R^n; and μ_k is the probability measure on B.

Under the unified framework of evolutionary computation, different random search algorithms are distinguished by their choices of neighborhood functions and selection functions. The design of the neighborhood function depends on the problem characteristics and the solution representation, for instance, crossover and mutation operators in genetic algorithms (GA) or random sampling in simulated annealing (SA). The selection function also varies by algorithm, such as the elitist selection strategy in GA or the Metropolis acceptance criterion in SA.

2.2 Formal Description of Basic PSO

In basic PSO, each particle represents a potential solution, and all particles form a swarm. Particles determine their velocity and direction of "flight" based on their own historical experience and social information from the swarm to search for the optimal solution.

2.2.1 Particle Representation

Consider a D-dimensional search space where the swarm consists of m particles, denoted as $Swarm = \{x_1^{(k)}, x_2^{(k)}, \cdots, x_m^{(k)}\}$. At time $k + 1$, the position vector of the i-th particle is $x_i^{(k+1)} = (x_{i1}^{(k+1)}, x_{i2}^{(k+1)}, \cdots, x_{iD}^{(k+1)})$, where $i = 1, 2, ..., m$. This position vector is the particle's location in the search space and also represents a potential solution to the problem. Associated with this position vector is the velocity vector $v_i^{(k+1)} = (v_{i1}^{(k+1)}, v_{i2}^{(k+1)}, \cdots, v_{iD}^{(k+1)})$, , which describes the particle's motion in each dimension of the space. (Note: Superscripts in parentheses indicate the iteration cycle, e.g., $x_{id}^{(k+n)}$ represents the position at the $(k + n)$ cycle; variables without parentheses represent powers, e.g., ω^n represents ω raised to the n-th power.)

2.2.2 *Neighborhood Function in PSO*

The neighborhood function in PSO generates new position states in each iteration by integrating the particle's position, velocity, personal experience (*pbest*), swarm information (*gbest*), and perturbations. For the d-th dimension of the i-th particle at time $k + 1$, the standard PSO neighborhood function is defined as:

$$\begin{cases} v_{id}^{(k+1)} = \omega \cdot v_{id}^{(k)} + c_1 \cdot r_1 \cdot (p_{id}^{(k)} - x_{id}^{(k)}) + c_2 \cdot r_2 \cdot (p_{ld}^{(k)} - x_{id}^{(k)}) \\ x_{id}^{(k+1)} = x_{id}^{(k)} + v_{id}^{(k+1)} \end{cases} \tag{2.1}$$

When generating a new particle vector, the velocity vector must satisfy the following constraint:

$$\left| v_{id}^{(k+1)} \right| \le V_{\max} \tag{2.2}$$

The constraint Eq. (2.2) can also be expressed as: $\left| x_{id}^{(k+1)} - x_{id}^{(k)} \right| \le V_{\max}$. It is a specific form of the Lipschitz condition for dynamic systems, ensuring bounded motion in each dimension.

Although the Lipschitz condition limits the magnitude of the velocity vector during each iteration, the updated position vector might still exceed the search space S. If this occurs, $x_{id}^{(k+n)}$ must be adjusted using one of the following methods (illustrated in Fig. 2.1).

where

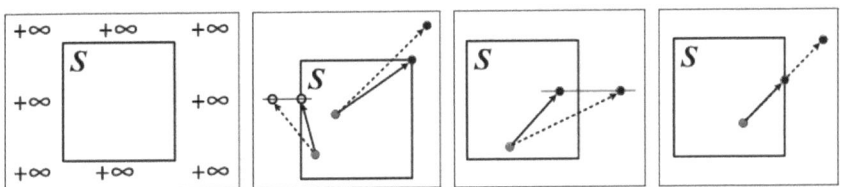

(*a*) Infinite boundary (absorbing walls)　(*b*) Nearest boundary (reflecting walls)　(*c*) Random reinitialization　(*d*) Scaling method (damping walls)

Fig. 2.1 Methods for particle resetting **a Infinite boundary (Absorbing Walls)**: For a minimization problem, if $x_i^{(k)} \notin S$, then set $f(x_i^{(k)}) = +\infty$ **b Nearest boundary (Reflecting Walls)**: If $x_i^{(k)} \notin S$, reset $x_i^{(k)} = x\prime$, where $x\prime$ is the closest point to $x_i^{(k)}$ within the search space S, satisfying $\forall x \in S, dist(x_i^{(k)}, x\prime) \le dist(x_i^{(k)}, x)$ **c Random Reinitialization**: Suppose the range of the d-th dimension of the search space S is $[lb_d, ub_d]$. If $x_i^{(k)} \notin [lb_d, ub_d]$, reset $x_i^{(k)} = lb_d + (ub_d - lb_d) \times r$, where r is a random number uniformly distributed within (0, 1) **d Scaling Method (Damping Walls)**: If $x_i^{(k)} \notin S$, reset $x_i^{(k)} = x_i^{(k-1)} + \sigma(x_i^{(k)} - x_i^{(k-1)})$,

Table 2.1 Parameters of the PSO algorithm

w	Inertia weight	V_{max}	Velocity limit
η	Velocity scaling factor	c_1, c_2	Acceleration factor
$x_{id}^{(k)}$	The current position vector of the particle	r_1, r_2	A random number between 0 and 1
$v_{id}^{(k)}$	The velocity vector of particle motion	m	Swarm size
$p_{id}^{(k)}$	The particle's personal best position	D	Search space dimension
$p_{ld}^{(k)}$	The swarm's best position	$f(\cdot)$	Fitness (objective) function

$$\sigma = \min_{\substack{\forall d, x_i^{(k)} \notin \\ [lb_d, ub_d]}} \left(\frac{||b_d - x_i^{(k-1)}||}{||x_i^{(k)} - x_i^{(k-1)}||} \right)$$

$$b_d = \begin{cases} ub_d, ub_d < x_i^{(k)} \\ lb_d, lb_d > x_i^{(k)} \end{cases}$$

2.2.3 Selection Function

The selection function in PSOupdates the personal best $p_{id}^{(k)}$ and global best $p_{ld}^{(k)}$ solutions::

$$p_{id}^{(k+1)} = \begin{cases} x_i^{(k+1)} & f(x_i^{(k+1)}) \leq f(p_{id}^{(k)}) \\ p_{id}^{(k)} & f(x_i^{(k+1)}) > f(p_{id}^{(k)}) \end{cases} \tag{2.3}$$

$$p_{id}^{(k)} \in \{x_{1d}^{(k)}, x_{2d}^{(k)}, \cdots, x_{md}^{(k)} | f(x_{id}^{(k)})\} = \min\{f(x_{1d}^{(k)}), f(x_{2d}^{(k)}), \cdots, f(x_{md}^{(k)})\} \tag{2.4}$$

$$p_{ld}^{(k)} \in \{p_{1d}^{(k)}, p_{2d}^{(k)}, \cdots, p_{md}^{(k)} | f(p_{id}^{(k)})\} = \min\{f(p_{1d}^{(k)}), f(p_{2d}^{(k)}), \cdots, f(p_{md}^{(k)})\} \tag{2.5}$$

The parameters in the above equations are defined in the Table 2.1.

2.2.4 Algorithm Steps

Steps of standard PSO are as follows:

Step 1: Initialize the swarm size m, inertia weight w, and acceleration coefficients c1, c2;

Step 2: Randomly initialize the position $x_{id}^{(k)}$ of each particle within the search space and initialize their velocity vectors $v_{id}^{(k)}$. Set the pbest $p_{id}^{(k)}$ for each particle to its current position, and calculate the gbest $p_{ld}^{(k)}$ using Eq. (2.5).

Step 3: Update the velocity using the neighborhood function Eq. (2.1); constrain velocities for each dimension according to Eq. (2.2); update the position of each particle by Eq. (2.1), and adjust infeasible positions using one of the methods in Fig. 2.1.

Step 4: Compute the objective function value for each particle, update *pbest* and *gbest* using Eqs. (2.4) and (2.5).

Step 5: If the maximum number of iterations is reached or the fitness error falls below a threshold, terminate and output the gbest; otherwise, return to Step 3.

2.2.5 Parameter Settings

Generally, standard PSO adopts real-number encoding and has relatively few parameters, typically set based on empirical evidence:

- **Swarm Size** (*m*): Usually set between 10 and 30, balancing computational efficiency and exploration. For most problems, a small number of particles is sufficient, which also helps reduce computational cost.
- **Velocity Limit** (V_{max}): Determines the maximum distance a particle can travel within a single iteration. Shi and Eberhart [7] provided simulation results for optimizing the PSO system, where V_{max} was set within $[2, X_{max}]$.
- **Acceleration Coefficients** (c_1 and c_2): Commonly set to 2.0. Clerc and Kennedy [8] recommends $(c_1 + c_2)/2 = 1.494$ for convergence.
- **Inertia Weight** (*w*): Typically is choosen between 0 and 1, with Clerc and Kennedy [8] selecting 0.729, Shi and Eberhart [7] suggesting values within [0.1, 1.05], and Riget and Vesterstroem [9] using a linearly decreasing schedule over time.
- **Velocity Scaling factor** (η): Usually set to 1.
- **Stopping Criteria**: Typically based on either maximum iterations or a predefined fitness tolerance (error threshold).

2.3 Mathematical Model of the PSO Algorithm

Since the introduction of PSO in 1995, researchers across disciplines have developed diverse algorithm models and formulations, analyzing and designing PSO from multiple perspectives. Pioneering contributions by Kennedy, Eberhart, Yuhui Shi, and Clerc established foundational models such as the basic PSO, inertia weight PSO, and constriction coefficient PSO and so on. Through extensive experiments, they systematically analyzed control parameters in these models and established corresponding reference values, inspiring subsequent enhancements by numerous scholars. The following sections detail key advancements in PSO modeling.

2.3.1 Inertia Weight PSO Model

In the earliest original PSO model, there was no inertia weight w. Shi and Eberhart [10] introduced inertia weight to dynamically balance exploration (global search) and exploitation (local refinement) capabilities, governed by the following equations:

$$\begin{cases} v_{id}^{(k+1)} = \omega \cdot v_{id}^{(k)} + c_1 \cdot r_1 \cdot (p_{id}^{(k)} - x_{id}^{(k)}) + c_2 \cdot r_2 \cdot (p_{ld}^{(k)} - x_{id}^{(k)}) \\ x_{id}^{(k+1)} = x_{id}^{(k)} + v_{id}^{(k+1)} \end{cases} \tag{2.6}$$

Their studies analyzed the impact of varying w on algorithm performance, including interactions with the velocity limit parameter V_{max}, unbounded velocity growth may cause swarm divergence. The introduction of w improve the balance between the swarm's exploration and exploitation. When $\omega > 1$, the velocity may increase over time, potentially leading the swarm to diverge. Conversely, when $\omega \leq 1$, induces velocity decay toward zero. A higher w promotes exploration of the search space and swarm diversity, while a lower w improves local exploitation. However, with excessively small w, the swarm may potentially stagnate in local optima.

The smaller the w, the greater the influence of the cognitive and social components on velocity updates. Particle updates depend on three factors: the previous velocity term, the "cognitive" component, and the "social" component. The inertia weight w modifies the influence of the previous velocity, while the "cognitive" and "social" components are influenced by the acceleration constants c_1 and c_2, respectively. Therefore, the relationship between the inertia weight and the acceleration constants is critical for parameter tuning.

2.3.2 Constriction Coefficient PSO Model

Clerc and Kennedy [8, 11] proposed scaling velocities via a constriction coefficient (χ) to ensure convergence without explicit velocity clamping. The mathematical formulation is as follows:

$$v_{id}^{(k+1)} = \chi \cdot (v_{id}^{(k)} + c_1 \cdot r_1 \cdot (p_{id}^{(k)} - x_{id}^{(k)}) + c_2 \cdot r_2 \cdot (p_{ld}^{(k)} - x_{id}^{(k)}))$$

$$x_{id}^{(k+1)} = x_{id}^{(k)} + \eta \cdot v_{id}^{(k+1)} \tag{2.7}$$

$$\chi = \frac{2\kappa}{|2 - \varphi - \sqrt{\varphi \cdot (\varphi - 4)}|}; \kappa \in [0, 1]; \varphi = c_1 + c_2$$

where κ controls the balance between the "exploration and exploitation" capabilities of the swarm: when $\kappa \approx 0$, the local exploitation capability leads to rapid convergence of the swarm, whereas when $\kappa \approx 1$, the algorithm is stronger in searching but slower in convergence. Typically, κ is assigned a fixed value. It is suggested to set $\varphi = 4.1$ (i.e., $c_1 = c_2 = 2.05$) and $\kappa = 1$, so derive $\chi \approx 0.729$. This is mathematically equivalent to setting the inertia weight $w = 0.729$ and $c_1 = c_2 = 1.49445$.

Introducing the constriction coefficient in PSO serves the same purpose as the introduction of inertia weight: balancing the trade-off between exploration and exploitation. For a specific value of χ, the equivalent model is $\omega = \chi$; $\varphi_1 = \chi \cdot c_1 \cdot r_1$; $\varphi_2 = \chi \cdot c_2 \cdot r_2$. The difference between these two methods is that the model with constriction coefficient does not require the use of Lipschitz conditions for velocity limiting, as it ensures convergence under specified constraints [12]. Clerc further refined this into a five-parameter time-invariant dynamic system, providing a general framework for analyzing algorithm stability.

2.3.3 Bare Bones Particle Swarm

Kennedy [13] proposed bare bones particle swarm (BBPS) replacing the standard velocity updates with Gaussian sampling around personal best (*pbest*) and global/neighborhood best (*gbest/lbest*). Furthermore, various swarm topologies and interaction probabilities of BBPS are also discussed. The specific equation is as follows:

$$x_{id}^{(k+1)} = N(\mu, \sigma^2)$$
$$\mu = (p_{ld}^{(k)} + p_{id}^{(k)})/2; \ \sigma^2 = |p_{ld}^{(k)} - p_{id}^{(k)}| \tag{2.8}$$

Poli and Langdon [14] analyzed BBPS as a discrete Markov chain using Cauchy and Gaussian distributions. He also noted that minor modification to the sampling distribution can significantly improve BBPS's search performance and even ensure its ability to reach the global optimum. Monson and Seppi [15] examined the optimization bias in BBPS. Pan et al. [16] demonstrated that BBPS can be derived from the standard PSO through mathematical deduction and discussed the "forgetting" characteristic of parameters in terms of probability.

Compared to standard PSO, BBPS is more streamlined and does not require parameter tuning. This makes it easier to implement and more compatible with other optimization algorithms. BBPS and its variations have been applied to integer programming [17] and image classification [18]. Omran et al. [19] utilized an embedded method to combine BBPS with the differential evolution (DE) algorithm, where the sampling points generated by BBPS serve as the base vectors for each update in DE.

2.3.4 Hybrid PSO Model

Hybrid PSO model was proposed by Angeline in 1998. The core concept integrates basic PSO with a selection mechanism. In HPSO model, new particles generated from each iteration are selected based on their fitness value. Specifically, the position and velocity vectors of the the top 50% of particles (with higher fitness) replace those of the bottom 50% (with lower fitness), while retaining the personal best positions (*pbest*) of the latter. This approach enhances convergence speed while preserving global search capabilities, achieving superior optimization results on most *Benchmark* functions compared to the original PSO model.

In 2000, *Mikkel Løvbjerg, Thomas K. Rasmussen*, and *Torben Krink* proposed further enhancing HPSO model by introducing crossover operations from evolutionary algorithms. The crossover mechanism operates as: (1) **Selection**: particles are selected for crossover with a predefined probability; (2) **Pairing and Crossover**: selected particles are randomly paired to perform crossover, generating "offspring". The position and velocity vectors of the offspring particles are determined as follows:

$$\text{child}_1(x_i^{(k)}) = \overrightarrow{p} \cdot \text{parent}_1(x_i^{(k)}) + (1 - \overrightarrow{p}) * \text{parent}_2(x_i^{(k)}) \tag{2.9}$$

$$\text{child}_2(x_i^{(k)}) = \overrightarrow{p} \cdot \text{parent}_2(x_i^{(k)}) + (1 - \overrightarrow{p}) * \text{parent}_1(x_i^{(k)}) \tag{2.10}$$

$$\text{child}_1(v_i^{(k)}) = \frac{\text{parent}_1(v_i^{(k)}) + \text{parent}_2(v_i^{(k)})}{|\text{parent}_1(v_i^{(k)}) + \text{parent}_2(v_i^{(k)})|} \cdot |\text{parent}_1(v_i^{(k)})| \tag{2.11}$$

$$\text{child}_2(v_i^{(k)}) = \frac{\text{parent}_1(v_i^{(k)}) + \text{parent}_2(v_i^{(k)})}{|\text{parent}_1(v_i^{(k)}) + \text{parent}_2(v_i^{(k)})|} \cdot |\text{parent}_2(v_i^{(k)})| \tag{2.12}$$

where $\text{child}_l(x_i^{(k)})$ and $\text{parent}_l(x_i^{(k)})$, $i = 1,2$ represent the position vectors of the offspring and parent particles, respectively. Similarly, $\text{child}_l(v_i^{(k)})$ and $\text{parent}_l(v_i^{(k)})$, $l = 1,2$ represent their velocity vectors, respectively. $\overrightarrow{p}$ is a random vector uniformly distributed across D dimensions, with each dimension taking values in the range [0,1].

The only difference between hybrid PSO and traditional PSO lies in the fact that hybrid PSO performs an additional crossover operation after updating velocities and positions. This crossover operation produces offspring particles that replace their parent particles. By inheriting the advantages of the parent particles, the offspring particles theoretically enhance the searching capability in the regions between particles. For instance, if two parent particles are located in different local optimum regions, their offspring produced through crossover can often escape from the local optima and achieve improved search results. Experimental results demonstrate that hybrid PSO exhibits faster searching speed and higher convergence accuracy compared to traditional PSO and genetic algorithms (GA). Currently, research into enhancing traditional PSO through evolutionary continues to advance.

2.3.5 P-Approximate Kalman Swarm (PAKS) Model

Monson and Seppi [20] proposed the P-Approximate Kalman Swarm (PAKS) model based on theoretical analysis and mathematical derivation of Bayesian optimization and Kalman filter approximations. The study compares the PAKS with PSO incorporated with inertia weight, constriction coefficient, and the noisy classical PSO through experimental simulations, thus demonstrating the effectiveness of the PAKS. The specific mathematical formulations are as follows:

$$
\begin{aligned}
\overline{v_{id}^{(k+1)}} &= (1-a) \cdot v_{id}^{(k)} + a \cdot b \cdot (p_{id}^{(k)} - x_{id}^{(k)}) + a \cdot (1-b) \cdot (p_{ld}^{(k)} - x_{id}^{(k)}) \\
v_{id}^{(k+1)} &\sim Normal(\overline{v_{id}^{(k+1)}}, \varphi \frac{||\overline{v_{id}^{(k+1)}}||^2}{D} I) \\
x_{id}^{(k+1)} &= x_{id}^{(k)} + v_{id}^{(k+1)}
\end{aligned}
\tag{2.13}
$$

2.3.6 Fully Informed Particle Swarm (FIPS) Model

References [21–23] discuss the Fully Informed Particle Swarm (FIPS), which modifies the heuristic and reference information in the particle update process. Unlike traditional models that solely rely on *gbest* and *pbest* information, FIPS employs an arithmetic mean based on $(p_{jd}^{(k)} - x_{id}^{(k)})$, where $p_{jd}^{(k)}$ comes from a neighbor j that is better than particle i at the current time step k. J represents the number of selected $p_{jd}^{(k)}$, which is not arbitrary [23]. The mathematical description of FIPS is as follows:

$$
\begin{aligned}
v_{id}^{(k+1)} &= x_{id}^{(k)} + w_1 \cdot (x_{id}^{(k)} - x_{id}^{(k-1)}) + Sum(r \cdot (w_2/J) \cdot (p_{jd}^{(k)} - x_{id}^{(k)})) \\
x_{id}^{(k+1)} &= x_{id}^{(k)} + v_{id}^{(k+1)}
\end{aligned}
\tag{2.14}
$$

2.3.7 Continuous PSO Model

Martínez [24] extended the standard PSO to a generalized GPSO framework and further derived a continuous PSO model based on a spring-damper system. The mathematical formulation is as follows:

$$
\begin{cases}
x''(t) + (1-\omega)x'(t) + \phi x(t) = \phi_1 p(t-t_0) + \phi_2 g(t-t_0), t \in R, \\
x(0) = x_0, \\
x'(0) = v_0.
\end{cases}
\tag{2.15}
$$

where $x(t)$ represents the trajectory of any particle in the swarm. Interaction with other particles is realized through the *lbest* $p(t)$ and *gbest* $g(t)$ (attractor). In this model, the average trajectory of the particles oscillates around a point:

$$o(t) = \frac{a_g g(t - t_0) + a_p p(t - t_0)}{a_g + a_l}. \tag{2.16}$$

Additionally, the attractor is delayed by t_0. In their work, this continuous model was applied to solve the parameters of an extreme learning machine (ELM) for neural networks (i.e., using the continuous PSO, specifically using the continuous PSO three-point model to address the inverse problem.). The trained ELM was then utilized to predict protein phosphorylation sites. Experimental results demonstrated the effectiveness of this model.

2.4 Topology of the PSO Algorithm

PSO is fundamentally rooted in the cooperative interactions among particles. This collective behavior emerges as particles exchange information with their neighboring particle and subsequently adjust their positions and velocities based on predefined strategies and received information. Such interactions lead to self-organization and drive the heuristic search process. The social network topology within PSO describes the neighborhood relationships and communication patterns among particles in the swarm. This topology governs information propagation in the swarm, thereby directly influencing the optimization capabilities and convergence properties. Current research on PSO topologies primarily focuses on two aspects:

(1) **Neighborhood Structure** (structural connectivity among particles);
(2) **Information Exchange Mechanisms** (the interaction pattern, e.g., communicate with global-best or local-best, or dynamic sharing rules).

Kennedy [13, 25] analyzed the topological structure of information transmission within the swarm and conducted seminal studies on PSO topologies. Currently, two primary methods exist for defining a particle's neighbors:

(1) **Index-Based Neighborhoods**, Neighbors are determined by particles with adjacent indices. In the index-based approach, particle positions are disregarded when determining neighbors. Particles are assigned fixed indices during initialization, resulting in their uchange neighbors relationship throughout optimization. This approach is predominantly used in **static neighborhood topologies**.
(2) **Position-Based Neighborhoods**. Neighbors are formed by particles with adjacent positions. The social network structure determines neighbors according to spatial proximity between particles, better preserving the swarm's geometric properties.. However, this method requires additional computational overhead

to calculate inter-particle distances and is typically employed in **dynamic neighborhood topologies**.

2.4.1 Static Neighborhood Topology

In static neighborhood topologies, neighborhood connectivity patterns remain fixed throughout optimization. Typically, a predetermined neighborhood index table is established during initialization based on particle indices. Among existing configurations, the fully-connected, ring, and Von Neumann structures are the most prevalent [21], as illustrated in Fig. 2.2.

Fully-connected Structure (Fig. 2.2a). Also termed the star topology, it connects all particles directly, forming mutual neighborhood relationships among all particles. This topology is commonly used in *gbest* PSO. All particles share the information of the best-performing particle in the swarm and are attracted to the *gbest*, accelerating convergence.

Ring Topology (Fig. 2.2b). It is the simplest neighborhood structure, where particles are directly connected to their closest k neighbors based on their indices. When $k = 2$, each particle has only two neighbors (left/right adjacency). When $k = N$-1 (where N = swarm size), the topology degenerates into a fully-connected structure. In this topology, updates depend solely on k immediate nearest neighbors, preserving swarm diversity.

Von Neumann Topology (Fig. 2.2c). It is another widely used neighborhood topology. Unlike the other two, the structure is a 2D grid where each particle is connected to four neighbors (above, below, left, and right). The Von Neumann structure has demonstrated robust performance in empirical study problem, particularly for high-dimensional optimization.

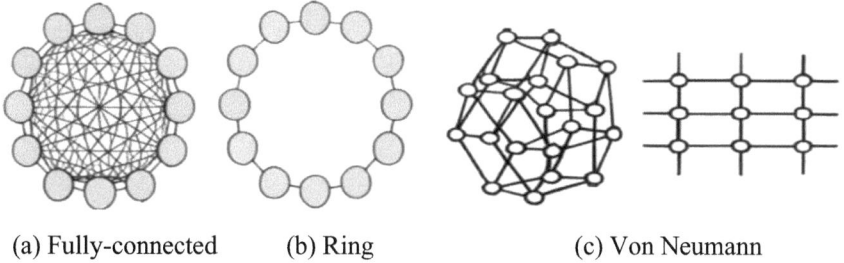

  (a) Fully-connected   (b) Ring   (c) Von Neumann

Fig. 2.2 PSO neighborhood topologies

2.4.1.1   Graph Properties of Static Neighborhood Topologies

Reference [26] analyzes the graph properties of the three static topologies discussed above, specifically calculating the diameter (DIA), average path length (APL), clustering coefficient (CC), and average degree (AD) of the three structures. For the fully connected structure, the graph properties are straightforward due to the direct connections between all nodes. In both the ring and Von Neumann structures, the AD represents the number of neighbors each node has; these structures do not have any nodes whose neighbors are also neighbors of each other, resulting in a CC of 0. These properties are summarized in Table 2.2.

Figure 2.3 illustrates how DIA and APL of the three structures change as the number of nodes N increases.

From Table 2.2 and Fig. 2.3, the following conclusions can be drawn:

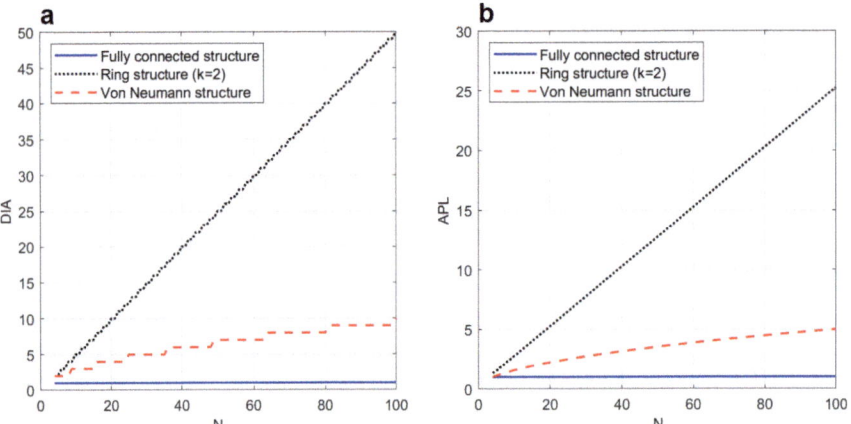

Fig. 2.3 Graph properties of static neighborhood topologies: **a** DIA; **b** APL

Table 2.2 Graph properties of fully connected, ring, and Von Neumann structures in PSO: neighborhood topologies and algorithm enhancements

	Diameter (*DIA*)	Average path length (*APL*)	Clustering coefficient (*CC*)	Average degree(*AD*)
Fully connected structure	1	1	1	N-1
Ring structure ($k = 2$)	$\lfloor N/2 \rfloor$	$\frac{N+1}{4}$ *or* $\frac{N^2}{4(N-1)}$ a	0	2
Von Neumann structure	$\left\lfloor \sqrt{N} \right\rfloor$ or $\sqrt{N} - 1$ b	$\approx \frac{\sqrt{N}}{2}$	0	4

Note

[a]For $N \geq 4$: when N is odd, APL = $(N + 1)/4$; When N is even, APL = $N^2/[4(N-1)]$.

[b]For $N \geq 16$: when N is odd, $DIA = \sqrt{N} - 1$; otherwise $DIA = \left\lfloor \sqrt{N} \right\rfloor$. $\lfloor A \rfloor$ denotes rounding down to the nearest integer.

(1) The fully connected structure has the smallest DIA, shortest APL, highest CC, and highest AD. The ring structure has the largest DIA, longest APL, lowest CC, and lowest AD. The Von Neumann structure has intermediate values for DIA, APL, and AD but has the smallest CC.

(2) The DIA and APL of the ring structure increase linearly with the number of nodes, with a relatively large rate of increase. In contrast, the diameter and APL of the Von Neumann structure increase with the square root of the number of nodes, but with a more gradual rate of increase.

(3) For all three structures, as AD increases, both DIA and APL decrease, though in a nonlinear manner. For example, the AD of the Von Neumann structure is twice that of the ring structure, but for the same number of nodes, its DIA and APL are significantly less than half of those of the ring structure.

(4) In the Von Neumann structure formed by different row and column configurations, variations in the number of rows and columns affect the graph properties.

2.4.1.2 Graph Properties and Performance Analysis

The graph properties of the three representative topologies have been described from the perspective of graph theory. Next, we proceed to analyze how these properties influence algorithm performance.

As shown in Table 2.2, the fully connected structure has the smallest DIA, shortest APL, and highest AD. These features allow any particle in the swarm to directly influence or be influenced by other particles. As a result, favorable information spreads rapidly throughout the swarm, which leads to fast convergence. However, this structure is prone to premature convergence or getting trapped in local optima. In the ring structure, nodes can only directly influence their immediate left and right neighbors and indirectly affect other particles through these neighbors. The larger DIA and APL of this structure mean that favorable information does not spread as quickly throughout the swarm. This indirect influence allows other particles to explore promising regions, thereby maintaining diversity within the swarm. Consequently, PSO based on the ring structure has strong optimization capabilities and is less likely to get trapped in local optima, but it converges more slowly. The Von Neumann structure features more balanced graph properties. Its smaller DIA and shorter APL allow favorable information to spread faster than in the ring structure while still not directly affecting all particles. Therefore, PSO based on the Von Neumann structure exhibits better convergence compared to the ring structure and superior optimization capabilities compared to the fully connected structure, making it a preferred topology [21, 27].

The relationship between structure and performance shows that DIA and APL influence the speed of information spread. A shorter DIA and APL facilitate faster dissemination and quicker convergence, but this can lead to premature convergence. Conversely, a longer DIA and APL result in slower convergence but reduce the risk of premature convergence. Moreover, DIA and APL are influenced by AD, which represents the average number of neighbors per particle. Typically, as AD increases (or decreases), both the diameter and average path length decrease (or increase). Consequently, some scholars argue that the average number of neighbors per particle is the most crucial factor affecting the neighborhood structure of particle swarms [23]. However, as discussed in Sect. 2.4.1.1, the relationship between AD, DIA, and APL is not strictly linear. In other words, structures with the same AD can exhibit significantly different DIA and APL. Additionally, CC differs between the ring and Von Neumann structures compared to the fully connected structure. This raises the question: What impact does the CC have on other graph properties, and how does it relate to algorithm performance? Next, let's examine the role of CC.

First, consider the process of adding neighbors to a node (particle). Due to the simplicity of the ring structure, we will investigate how nodes expand within this structure. As shown in Fig. 2.4a, node i can select different nodes as its neighbors. In Fig. 2.4b, node i chooses the neighbor of its current neighbor (nodes a and b) as new neighbors. In Fig. 2.4c, node i selects nodes further away (nodes c and d) as new neighbors. For node i, both methods of neighbor selection increase its degree, and the newly formed topologies have smaller DIA and APL compared to the original structure. However, the topology in Fig. 2.4b has a higher CC than the one in Fig. 2.4c. What differences arise from this variation? If we compare Figs. 2.4b and c, it is apparent that the new neighbors in Fig. 2.4c significantly shorten the distance between node i and other nodes, especially distant nodes, relative to those in Fig. 2.4b. This results in a smaller DIA and APL, as shown in Fig. 2.4c. A lower CC indicates that nodes tend to connect more to non-adjacent nodes when forming new edges, helping to reduce the distance to distant nodes. Therefore, under the same AD, structures with a lower CC are more likely to have smaller DIA and APL. Fig. 2.4d and e show two different topologies derived from the ring structure in Fig. 2.4a, where each node selects neighbors at distances of 4 and 2, respectively. Both topologies have the same AD ($AD = 4$), but the structure with the lower CC in Fig. 2.4d exhibits a smaller DIA and APL.

In real society, one characteristic of a group or faction is that information tends to be localized. Let's now examine how the formation of groups within a network affects the dissemination of information. In Fig. 2.4b, two neighbors of node e also form a connection and create a group. Suppose that node i currently outperforms node e, node a, and the other neighbors of node a. Node e can only gather external information by learning from its neighbors, node a and node i. Node a, in turn, adjusts itself by learning from its best neighbor, node i. Therefore, node e receives similar information from its neighbors and is forced to converge toward node i. Essentially, the neighbors of node e block its access to external information, making it difficult for node e to obtain new information and escape the region identified by node i. However, the situation is different for node e in Fig. 2.4c. Here, since node a is not directly influenced by node i, node e can receive more diverse information from its neighbors. Thus, when neighbors of a node are more evenly distributed across the network, the node can gather more comprehensive information, making it less likely to become trapped in a local area. CC can help regulate the distribution of neighbors in the network. Interestingly, based on certain construction rules of graphs (e.g., *CC*

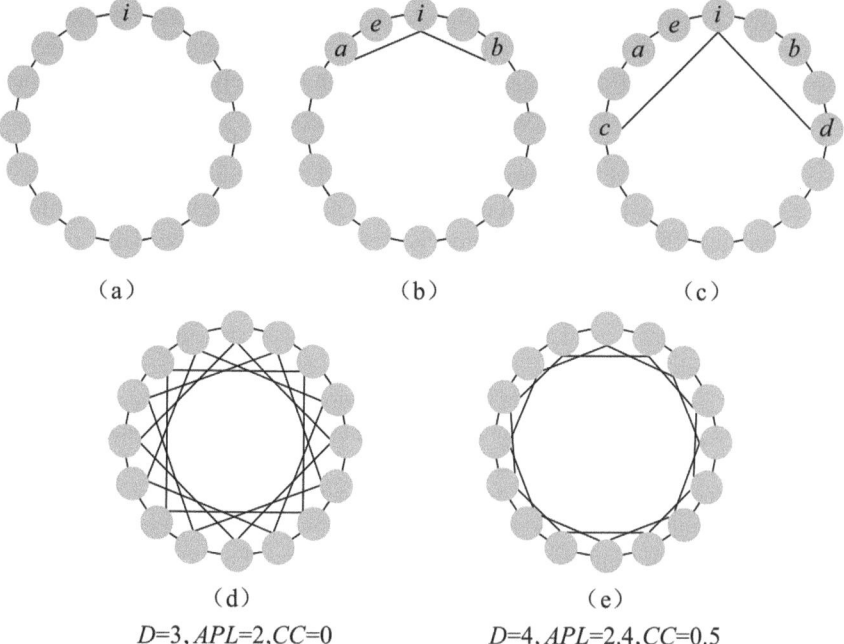

Fig. 2.4 Ring structure and its variants

$= 0$ and $AD = 4$), adding edges to a ring structure can form a network similar to the Von Neumann structure. For instance, in Fig. 2.4d, each node has 4 neighbors, forming a grid with properties similar to a 16-node Von Neumann structure. In fact, starting from a ring structure, we can create more flexible and efficient topologies than the Von Neumann structure by connecting nodes at varying distances.

From the above analysis, we can conclude that CC significantly influences the structure, thereby affecting PSO performance. When AD (i.e., the average number of neighbors) remains constant, a structure with a lower CC not only has a smaller DIA and shorter APL but also more evenly distributed neighbors. This helps algorithms based on such structures avoid becoming trapped in local optima and achieve faster convergence.

2.4.2   Dynamic Neighborhood Topology

Unlike static neighborhood topologies, PSO based on dynamic neighborhood topologies can adjust the neighbor structure or information-sharing mechanisms in real time according to the current state of the particles. Neighbor formation based on positional proximity is often used to generate dynamic topologies [28–30]. However, this approach requires calculating the distances between particles, which increases computational load as the dimensionality rises. Furthermore, distance-based neighborhood structures are prone to forming isolated local regions with limited interaction between particles in different regions, making them more likely to be trapped in local optima [28, 29] or exhibit slower convergence [30].

In reference [23], c_1 and c_2 were set to 0, resulting in the "social-only model" ($c_1 = 0$), the "cognitive-only model" ($c_2 = 0$) and the "selfless mode" (where $c_1 = 0$ and a particle's pbest solution is excluded from its neighborhood), with an analysis of the performance of these models. Reference [31] concluded that a smaller neighborhood structure tends to perform better in solving complex problems, while a larger neighborhood structure excels in simpler problems. In reference [32], the swarm topology was extended into a random pseudo-star structure by introducing crossover operations.

Reference [33] designed a dynamic neighborhood topology where each particle's neighborhood gradually expands until it encompasses all particles in the swarm. In reference [34], a dynamic hierarchical structure was constructed to enable different information (neighborhood) structures. Based on the optimal fitness value of the most recent iteration, particles are moved up or down a level, where higher-level particles exert greater influence on the rest of the swarm. Furthermore, at different levels, particles are assigned distinct behaviors. The study examined both fixed tree-like structures and dynamic hierarchical structures. It also investigated the use of different inertia weights ω for particles at different levels in the dynamic hierarchical structure. The numerical simulation indicated that the algorithm for dynamic hierarchical structures outperformed that of the fixed tree-like structures, and incorporating ω sped up convergence compared to the purely dynamic hierarchical approach.

A dynamic neighborhood topology was also designed in reference [35], where, in each iteration cycle, a particle selects m particles with the closest fitness values to form its new neighborhood. Based on this concept, a PSO was developed that integrates fitness function, distance, and proportion. When updating each dimension of the velocity vector, the algorithm selects other particles with higher fitness and closer proximity for the update.

In addition to studying topologies formed by information exchange within individual swarms, many scholars have explored information exchange and topologies in multi-swarm PSO. In reference [36], a method for exchanging information between multiple swarms was proposed. By exchanging individual particles between swarms (similar to multi-population genetic algorithms), each swarm is divided into two parts: one for exchange and the other to be replaced by particles from other swarms. Hamming distance is used to measure the relationship between a swarm and its neighbors when selecting exchanging neighbors. The introduction of neighbor best in velocity updates allows each particle to be influenced by the pbest, gbest, and neighborhood best during each iteration.

For algorithms characterized by different swarms, *Ray* designed a model incorporating social and civilization aspects [24], where individuals from different societies collaborate to enhance their performance. This model features two levels of migration: intra-society, where individuals move toward the leader within their own society, and inter-society, where leaders from different societies are attracted to leaders of a higher level of civilization.

To enhance the topology during information transmission and increase diversity in optimization, Al-Kazemi and Mohan [37] split the swarm into two groups randomly. One group follows the motion of *gbest*, while the other group moves in the opposite direction. After a certain period, if *gbest* shows no improvement, particles will actively switch between the two groups.

Baskar and Suganthan [38] constructed two groups that frequently exchange the *gbest* information within their respective swarms, with both swarm searching simultaneously. After each exchange of information, both groups track the better *gbest*. One group updates using standard PSO, while the other group uses the "fitness-distance ratio". Simulation results indicate that this method outperforms the two methods used independently in solving single-objective optimizations.

El-Abd and Kamel [39] further improved the method by introducing a bidirectional information flow between the two groups. Specifically, the best particles from each group are exchanged to replace the worst particles in the other group. After a certain number of iterations, the top *p* best particles from one group will replace the worst *p* particles in the other group, provided that the fitness of these *p* best particles is higher than that of the worst *p* particles in the other group. This mechanism allows the two groups to continuously acquire new information from each other.

Parspopoulos [40] designed the vector evaluated particle swarm optimizer (VEPSO), a multi-swarm structure in which each swarm optimizes one component of the objective function. The optimized information is also exchanged with other swarms that optimize different objective functions. In VEPSO, the gbest used to update particle velocities within each group is selected from other groups.

2.5 Evaluation Metrics for PSO

2.5.1 Accuracy

Accuracy refers to the quality of the solution obtained by the algorithm, specifically how well the PSO identifies the *gbest*. If the fitness of the optimal solution $f(x^*)$ is known, accuracy can be defined as the absolute difference between the fitness $f(x^*)$ of the known optimal solution and the fitness $f(x)$ of the solution found by the algorithm.

$$\text{accuracy}(S, t) = |f(x) - f(x^*)| \qquad (2.17)$$

When there is no prior information about the optimal solution, the accuracy at any given time t is represented by the fitness value of *gbest* at that moment.

$$\text{accuracy}(S, t) = f(x) \qquad (2.18)$$

2.5.2 Reliability

Reliability refers to the proportion of experiments where the algorithm achieves a specified level of accuracy (fitness or error) relative to the total number of experiments conducted. The reliability of a swarm can be defined as:

$$\text{reliability}(S, \varepsilon) = \frac{n}{N} \cdot 100\% \tag{2.19}$$

where ε is the specified accuracy level, N is the total number of simulations, and n is the number of simulations that converge to the specified accuracy. A higher value of $reliability(S, \varepsilon)$ indicates greater reliability of the algorithm.

2.5.3 Robustness

Robustness, or stability, refers to the consistency of the algorithm's performance across multiple simulations. The less variation in performance, the more robust the algorithm is considered to be.

2.5.4 Diversity

Diversity in a swarm refers to the variability among the individuals within that swarm. For swarm-based optimization algorithms, high diversity is crucial as it enables exploration of a larger area of the search space. In simple terms, diversity can be defined as the degree of dispersion among particles. The concept of swarm diversity will be discussed in detail in Sect. 2.6.

2.6 Study on Diversity

In the later stages of evolution, PSO may lead to a swarm where many particles become similar (i.e., particles converge to a very small region of the search space, leaving other regions unoccupied). As a result, the swarm may fail to evolve better particles. When diversity decreases, the search can become trapped in a region that does not contain gbest and is difficult to escape, thus causing the algorithm to be stuck in local optima. Therefore, maintaining diversity allows the algorithm to continue exploring unknown regions, ensuring the global search capabilities and the optimality of the final results.

To avoid premature convergence, some scholars have proposed to enhance overall performance by controlling swarm diversity. Jacques and Jako [41] designed a method based on standard PSO, which controls swarm characteristics through diversity metrics, thereby achieving a balance between attraction and repulsion among particles to avoid premature convergence. This method introduces a repulsion phase after the attraction phase of position updates, i.e., the inverse of attraction phase. This inverse partially suppresses the decrease in system diversity caused by the attraction phase. If the system diversity drops below a predetermined threshold, the algorithm switches to the repulsion phase to increase diversity. When the repulsion phase restores this diversity to the predetermined level, repulsion ends, and the algorithm continues with the basic algorithm.

Morter and Krink [42] proposed another approach to improve diversity—self-organized criticality (SOC). Similarly, the fundamental idea of SOC is to increase particle diversity during the search. The only difference between SOC particles and standard particles is that each SOC particle includes a critical value (C), which is initialized to zero. The closer the particles are, the poorer the diversity. Therefore, if the distance between two particles is less than a predetermined threshold, their critical values are increased. When a critical value exceeds the global limit, its position in the solution space is reallocated, thus effectively enhancing the diversity of the search.

Krink [43] also addressed the diversity issue by proposing SPEPSO, which expands particle space to solve problems related to particle collisions and clustering, as well as enhancing their ability to escape local minima. Each particle is assigned a minimum independent radius r. If the distance between two particles is less than r, it is considered that they are in "friction," and control measures are applied to separate them. This algorithm also addresses how to determine the direction and speed of particle separation, and offers three main strategies: random rebound, real physical rebound, and rebound along the original trajectory. Moreover, a rebound velocity factor (ranging between 0 and 1) is introduced to adjust the original velocity and prevent particle collisions. Buthainah Al-Kazzemi et al. [44] proposed another method to enhance the diversity by setting different objectives at various phases of the optimization. Particles are driven to approach or move away from their known *pbest* or *gbest*, depending on the phase. It exhibited superior performance compared to standard PSO in both discrete and continuous optimizations, with higher computational efficiency than standard PSO or genetic algorithms.

Particle diversity is a key measure in PSO, reflecting how the particles are distributed in the search space. It plays a crucial role in the algorithm's performance. When particles (or their velocities and pbest) become too close to each other, the algorithm is likely only performing a local search, potentially missing out on other regions of the search space. This can result in missing the gbest, which negatively impacts the algorithm's effectiveness. Therefore, studying and maintaining particle diversity is critical.

2.6.1 Definition of Diversity

Diversity is defined as a measure of the search process in evolutionary algorithms. In general, diversity does not evaluate whether the algorithm has found the optimal solution but rather evaluates the distribution of individuals (current solutions) within the population. In PSO, there are multiple ways to measure diversity. Shi and Eberhart [45] introduced two different measures of diversity to assess the search progress in PSO.

(1) Position Diversity of the Swarm

A. Position Diversity at the Element Level

Based on reference [46, 47] concerning swarm diversity in genetic algorithms, the elements in each dimension are measured uniformly. The position diversity of the swarm at the element level is defined as follows:

$$\bar{x} = \frac{1}{m \times n} \sum_{i=1}^{m} \sum_{j=1}^{n} x_{ij}$$

$$D^{p} = \frac{1}{m \times n} \sum_{i=1}^{m} \sum_{j=1}^{n} [x_{ij} - \bar{x}]^{2}$$

(2.20)

where m represents the swarm size, n represents the dimensionality of the particles (i.e., the number of variables), the superscript p in D^{p} represents the position diversity of the swarm. If the superscript is v, then it indicates the velocity diversity.

B. Position Diversity based on Euclidean Distance

To measure swarm diversity, the Euclidean distance between every pair of particles is calculated. The corresponding equations are as follows:

$$d^{p}(x_i, x_j) = ||x_i - x_j||$$

$$d^{p}(x_i, x_j) = \frac{d^{p}(x_i, x_j)}{||ub - lb||}$$

$$D_{ED}^{p} = \frac{2}{m(m-1)} \sum_{i=1}^{m} \sum_{j=i+1}^{m} d^{p}(x_i, x_j)$$

(2.21)

where $[lb, ub]$ represents the dynamic range of particle positions, or the search space of the particles. References [48, 49] have explored this type of diversity metric in detail.

C. Dimension-wise Position Diversity

In addition to evaluating whether particles are close to each other based on a specific distance, swarm diversity can also be assessed by considering the position of particles in each dimension., thereby measuring the diversity of positions across each dimension. The calculation is expressed in Eq. (2.22) where m is the swarm size, and n is the number of dimensions.

$$\bar{x}_j = \frac{1}{m} \sum_{i=1}^{m} x_{ij}$$

$$D_j^p = \frac{1}{m} \sum_{i=1}^{m} [x_{ij} - \bar{x}_j]^2$$

(2.22)

Thus, the dimension-wise position diversity vector $D^v = [D_1^v, ..., D_n^v]$ provides a dimension-based diversity measurement approach, and there exist multiple methods available for calculating the position diversity of the swarm.

① **Weighted Position Diversity**

Different parameters (i.e., different dimensions) have different weights. When calculating position diversity, weights are assigned to each element of $D^v = [D_1^v, ..., D_n^v]$. The equation for calculating the position diversity is as follows:

$$D_{WS}^p = \sum_{j=1}^{n} w_j D_j^p$$

(2.23)

Each w_j is a positive real number less than or equal to 1. iFeach dimension contributes equally, then $w_j = 1/n$. The above equation can be rewritten accordingly:

$$D_{WS}^p = \frac{1}{n} \sum_{j=1}^{n} D_j^p$$

(2.24)

② **Maximum Weighted Position Diversity**

First, weights are assigned to each dimension of position diversity, and then the maximum value is calculated. The equation is as follows:

$$D_{WS}^p = \max\{w_j D_j^p\} \quad j = 1, ..., n$$

(2.25)

③ **Vector norm of Position Diversity**

The vector norm of dimension-wise position diversity is calculated based on Euclidean distance. The equation is as follows:

$$D_L^p = \sqrt{\sum_{j=1}^{n} D_j^{p2}}$$

(2.26)

④ **Normalization of the Dimension-wise Position Diversity**

In most PSO algorithms, the search range is constrained within a specific search space. Assuming the dynamic search space of the j-th dimension of the position matrix X is $[lb_j, ub_j]$, the normalization equation for the position X is shown in Eq. (2.27).

$$x_{ij} = \frac{x_{ij}}{|ub_j - lb_j|} \qquad lb_j \le x_{ij} \le ub_j;$$
$$i = 1, ..., m; \qquad j = 1, ..., n \tag{2.27}$$

The normalization method for the dimension-wise position diversity is as follows:

$$\bar{x}_j = \frac{1}{m} \sum_{i=1}^{m} x_{ij}$$
$$D_{jN}^p = \frac{1}{m} \sum_{i=1}^{m} \left[x_{ij} - \bar{x}_j \right]^2 \tag{2.28}$$

After the matrix is normalized, the position diversity is calculated using weighted sum, maximum weight, or the vector norm.

(2). **Velocity Diversity of the Swarm**

A distinguishing feature of the PSO algorithm compared to other evolutionary algorithms is that, in addition to the position matrix, there is also a velocity matrix. Thus, when studying swarm diversity, it is essential to consider not only position diversity but also velocity diversity. The particle's velocity comprises two aspects: magnitude (speed) and direction. Consequently, examining particle velocity requires addressing both of them: **Quantifying Magnitude Diversity** (calculating speed diversity) and **Quantifying Directional Diversity** (calculating diversity in velocity direction). Since particles move along dimensions, the magnitude diversity should be calculated using intra-dimensional movement. For directional diversity the calculation is based on the angle between the velocity vectors of all particles and the swarm's average velocity vector.

A. **Velocity Magnitude Diversity in the Space**

To calculate the magnitude diversity, the average magnitude in the j-th dimension is obtained by averaging the magnitudes of all particles in that dimension. The equation for dimension-wise magnitude diversity is as follows:

$$\bar{v}_j = \frac{1}{m} \sum_{i=1}^{m} v_{ij}$$
$$D_j^{v\_ds} = \frac{1}{m} \sum_{i=1}^{m} [v_{ij} - \bar{v}_j]^2 \tag{2.29}$$

where $D_j^{v\_ds}$ represents the magnitude diversity in the j-th dimension.

Similar to dimension-wise position diversity, its calculation includes: weighted sum of magnitude diversity, maximum weighted magnitude diversity, and the vector norm of magnitude diversity. The corresponding formulas are provided in Eqs. (2.23), (2.24), (2.25) and (2.26).

B. Directional Diversity

Directional diversity measures the diversity of particles based on the direction of their velocity vectors. To calculate directional diversity, the first step is to normalize the elements of each particle's velocity vector:

$$v_{ij}^{nor\_par} = \frac{v_{ij}}{\sqrt{\sum_{j=1}^{n} v_{ij}^2}} \tag{2.30}$$

Next, the average velocity vector of the particles is normalized. First, compute the average for each dimension of the velocity vectors, as shown in Eq. (2.31). Then, normalize each dimension of average velocity vector to obtain the normalized average velocity vector, as below::

$$v_j^{dim\_avg} = \frac{1}{m} \sum_{i=1}^{m} v_{ij} \tag{2.31}$$

$$v_j^{nor\_avg} = \frac{v_j^{dim\_avg}}{\sqrt{\sum_{j=1}^{n} v_j^{dim\_avg2}}} \tag{2.32}$$

Finally, compute the cosine of the angle θ between each particle's normalized velocity vector and the normalized average velocity vector, as shown in Eq. (2.33). The directional diversity of the swarm is calculated using Eq. (2.34).

$$\cos(\theta_i) = \sum_{j=1}^{n} (v_{ij}^{nor\_par} \cdot v_j^{avg\_nor}) \tag{2.33}$$

$$D^{v\_dir} = \frac{1}{m} \sum_{i=1}^{m} \theta_i^2 \tag{2.34}$$

2.6.2   Normalization of Swarm Diversity

In PSO research, the problems addressed often differ substantially, and varying dynamic ranges correspond to different problems. Consequently, the dynamic range of swarm diversity also varies across different problems, making it necessary to use distinct diversity calculation methods for different scenarios. Therefore, defining a normalized diversity metric becomes critically important. *Cheng* and *Shi* [50] proposed a method for normalizing swarm diversity.

Optimization problems are generally categorized into separable and non-separable ones. In separable problems, each x_{ij} contributes independently to the fitness value of x_i, whereas whereas in non-separable problems, the contribution is interdependent. Therefore, for separable optimizations, swarm diversity is typically calculated using a dimension-wise swarm diversity method, while for non-separable problems, an element-wise approach is employed.

(1)   Vector Norms and Matrix Norms

A vector norm is a function that maps a vector space $\mathbb{R}^n$ to $\mathbb{R}$: $f : \mathbb{R}^n \rightarrow \mathbb{R}$. In finite-dimensional space, all norms are equivalent. For example, if $|| \cdot ||_\alpha$ and $|| \cdot ||_\beta$ are norms in space $\mathbb{R}^n$, there exist positive constants c_1 and c_2 such that $c_1 ||x||_\alpha \leq ||x||_\beta \cdots \leq c_2 ||x||_\alpha, \forall x \in \mathbb{R}^n$. The vector norms have the following properties:

$$||x||_1 \geq ||x||_2 \cdots \geq ||x||_\infty \tag{2.35}$$

Similarly, a matrix norm is a function that maps the space $\mathbb{R}^{m \times n}$ to $\mathbb{R}$: $f : \mathbb{R}^{m \times n} \rightarrow \mathbb{R}$. The matrix norms have the following properties:

$$||A||_1 = \max_{1 \leq j \leq n} \sum_{i=1}^{m} |a_{ij}|, ||A||_\infty = \max_{1 \leq i \leq n} \sum_{j=1}^{n} |a_{ij}|.$$

In the context of applying matrix norms to PSO matrix norms are interpreted as follows:

- **For each dimension**, compute the sum of absolute values of the elements in each column vector of the position matrix, and take the maximum value as the L_1 norm of the position matrix.
- **For each particle**, compute the sum of absolute values of the elements in its position vector, and take the maximum value as the L_∞ norm of the position matrix.
- The distinction between the L_1 and L_∞ norms lies in the aspect of the position matrix they measure:
- The **L_1 norm** measures the column sums of the position matrix (aggregated per dimension).

- The L_∞ **norm** measures the row sums of the position matrix (aggregated per particle)..
- Based on the properties of norms:

For **separable problems**, vector norms can be applied to normalize swarm diversity, regardless of vector independence.

For **non-separable problems**, the L_∞ norm is often used to normalize swarm diversity.

(2) Normalization of Swarm Diversity

A. Normalization of Position Diversity

Position diversity reflects the distribution of particles' current positions. By measuring position diversity, we can determine whether the swarm is diverging into a broader search space or converging into a smaller region. Position diversity typically focuses on the vector elements within the position matrix.

Separable problems: For separable problems, each vector in the position matrix is independent. Therefore, vector norms are used to normalize positions, and there are three normalization methods. These methods are based on the L_1 norm, L_∞ norm, and the normalization based on the maximum value of the position:

$$x_{ij}^{nor} = x_{ij}/||\mathbf{x}||_1 = x_{ij}/\sum_{j=1}^{n}|x_{ij}|, \ x_{ij}^{nor} = x_{ij}/||\mathbf{x}||_\infty$$

$$= x_{ij}/\max|x_{ij}| \ and \ x_{ij}^{nor} = x_{ij}/X_{max}$$

The normalized position diversity for separable problems is calculated as follows:

$$\overline{\mathbf{x}}^{nor} = \frac{1}{m}\sum_{i=1}^{m}x_{ij}^{nor}, \ D^p = \frac{1}{m}\sum_{i=1}^{m}\left|x_{ij}^{nor} - \overline{x}_j^{nor}\right| \ and \ D^p = \frac{1}{n}\sum_{j=1}^{n}D_j^p$$

where $D^p = [D_1^p, ..., D_n^p]$ represents the diversity of each dimension, and D^p is the normalized position diversity of the particles in the swarm.

Non-separable problems: For non-separable problems, vectors in the position matrix are interdependent, and this interdependence must be considered when calculating diversity. Based on $|x_{ij}|_{max}$, L_∞ norm, or the maximum position value in the matrix X, there are also three normalization method for each particle's position:

$$x_{ij}^{nor} = x_{ij}/\max|x_{ij}|, \ x_{ij}^{nor} = x_{ij}/||X||_\infty = x_{ij}/\max_{1\le i\le m}\sum_{j=1}^{n}|x_{ij}| \ and \ x_{ij}^{nor} = x_{ij}/X_{max}$$

After normalizing positions, the normalized position diversity for non-separable problems is as follows:

$$\bar{x}^{nor} = \frac{1}{m \times n} \sum_{i=1}^{m} \sum_{j=1}^{n} x_{ij}^{nor} \; and \; D^p = \frac{1}{m \times n} \sum_{i=1}^{m} \sum_{j=1}^{n} |x_{ij}^{nor} - \bar{x}^{nor}|$$

where D^p is the normalized position diversity of the particles in the current iteration.

B. Normalization of Velocity Diversity

Velocity diversity reflects the movement trends of particles, indicating the distribution characteristics of their current velocities. In other words, velocity diversity measures the activeness of particles. By examining velocity diversity, we can infer whether the swarm is diverging or converging. Particle velocity consists of two components: magnitude and direction. Since direction diversity does not require normalization, only the magnitude of velocity diversity is normalized here.

Separable problems: For separable problems, vectors in the velocity matrix are independent. Velocity normalization is conducted using the L_1 norm, L_∞ norm, and the maximum value of the velocity:

$$v_{ij}^{nor} = v_{ij}/\|v\|_1 = v_{ij}/\sum_{j=1}^{n} |v_{ij}|, \; v_{ij}^{nor} = v_{ij}/\|v\|_\infty = v_{ij}/\max |v_{ij}| \; and$$

$$v_{ij}^{nor} = v_{ij}/V_{\max}.$$

The normalized velocity diversity for separable problems is calculated as:

$$\bar{v}^{nor} = \frac{1}{m} \sum_{i=1}^{m} v_{ij}^{nor}, \; D^v = \frac{1}{m} \sum_{i=1}^{m} |v_{ij}^{nor} - \bar{v}_j^{nor}| \; and \; D^v = \frac{1}{n} \sum_{j=1}^{n} D_j^v.$$

where $D^v = [D_1^v, ..., D_n^v]$ represents the diversity of each dimension, and D^v is the normalized velocity diversity for each particle.

Non-separable problems: For non-separable problems, vectors in the velocity matrix are interdependent. There are three methods for normalizing velocity diversity: $|v_{ij}|_{\max}$ of the velocity matrix, L_∞ norm of the V matrix, and the maximum value of the velocity. That is:

$$v_{ij}^{nor} = v_{ij}/\max |v_{ij}|, \; v_{ij}^{nor} = v_{ij}/\|V\|_\infty = v_{ij}/\max_{1 \le i \le m} \sum_{j=1}^{n} |v_{ij}|, \; or \; v_{ij}^{nor} = v_{ij}/V_{\max}.$$

The method for normalizing velocity diversity in non-separable problems is:

$$\bar{v}^{nor} = \frac{1}{m \times n} \sum_{i=1}^{m} \sum_{j=1}^{n} v_j^{nor} \; and \; D^v = \frac{1}{m \times n} \sum_{i=1}^{m} \sum_{j=1}^{n} |v_{ij}^{nor} - \bar{v}^{nor}|$$

where D^v represents the normalized velocity diversity of each particle in the current iteration.

C. **Cognitive Diversity**

Cognitive diversity reflects the distribution of each particles' current personal best positions *pbest*, which partially traces the algorithm's search trajectory toward the optimal solution. The calculation for cognitive diversity is similar to that of position diversity, except uses the current *pbest*. Therefore, the analytical framework for position diversity is equally applicable to cognitive diversity.

Separable problems: The normalization for cognitive position is:

$$p_{ij}^{nor} = p_{ij}/\|p\|_1 = p_{ij}/\sum_{j=1}^{n}|p_{ij}|, p_{ij}^{nor} = p_{ij}/\|p\|_\infty = p_{ij}/\max|p_{ij}|, \text{ and}$$

$$p_{ij}^{nor} = p_{ij}/X_{\max}$$

The normalization cognitive diversity for separable problems is as follows:

$$\overline{p}^{nor} = \frac{1}{m}\sum_{i=1}^{m}v_{ij}^{nor}, D^c = \frac{1}{m}\sum_{i=1}^{m}|v_{ij}^{nor} - \overline{v}_j^{nor}|, \text{ and } D^p = \frac{1}{n}\sum_{j=1}^{n}D_j^v$$

where $D^c = [D_1^c, ..., D_n^c]$ represents the diversity of each dimension, and D^c is the normalized cognitive diversity of each particle.

Non-separable problems: Similar toposition diversity, the normalization method for a particles' *pbest* is:

$$p_{ij}^{nor} = p_{ij}/\max_{i,j}|p_{ij}|, p_{ij}^{nor} = p_{ij}/\|P\|_\infty = p_{ij}/\max_{1\le i\le m}\sum_{j=1}^{n}|p_{ij}| \text{ and } p_{ij}^{nor} = p_{ij}/X_{\max}.$$

The normalization cognitive diversity for non-separable problems is:

$$\overline{p}^{nor} = \frac{1}{m\times n}\sum_{i=1}^{m}\sum_{j=1}^{n}p_j^{nor} \text{ and } D^c = \frac{1}{m\times n}\sum_{i=1}^{m}\sum_{j=1}^{n}|p_{ij}^{nor} - \overline{p}^{nor}|$$

where D^c is the normalized cognitive diversity of a particle during the current iteration.

Performance Evaluation in PSO

When assessing PSO performance, one may separately study position, speed, direction, and cognitive diversity:

Position diversity measures the distribution of particles' current positions. Observing it reveals whether the swarm is diverging into a broader search space or converging to a smaller region.

Speed and direction diversity of velocity provide insights into the particles' movement trends quantifying the distribution characteristics of their current velocities. Magnitude diversity indicates divergence/convergence tendencies, while directional diversity reflects the activeness of particle motion.

Cognitive diversity captures the distribution of each particle's current *pbest*, tracing the algorithm's search trajectory toward optimal solution.

In addition to analyzing these diversity measures separately, since position, magnitude, and directional diversity all describe particle movements, it is also possible to consider them collectively rather than focusing on just one or two of them. Equation (2.36) exemplifies a method for calculating overall diversity.

$$D = w_p D^p + w_{v\_ds} D^{v\_ds} + w_{v\_dir} D^{v\_dir} \tag{2.36}$$

Swarm diversity quantifies the distribution of particles in the search space and tracks the algorithm's exploration progress, which is critical for PSO research. Moreover, the dynamic range of diversity varies across optimization problems due to differing problem-specific characteristics. Therefore, diversity calculations must be tailored to the specific problem to enable accurate analysis.

2.7   Premature Convergence in PSO

A key issue in evolutionary algorithms is premature convergence [9, 51], which occurs during the search process in complex solution spaces. This issue can lead to suboptimal solutions and degraded optimization performance. In genetic algorithms, excessive selection pressure and high gene flow among individual particles in the swarm are major causes of premature convergence. In PSO, excessive information exchange between particles and a lack of diversity in information flow can accelerate particle clustering. As swarm diversity decreases, the algorithm becomes less capable of escaping local optima.

Like other evolutionary algorithms, PSO also faces the problem of premature convergence. Various methods have been proposed in the literature to address this. These include introducing an additional "Queen" particle into the swarm [11], modifying the topology of information exchange between particles [52], expanding the swarm by adding subpopulation [53], Bergh's introduction of a random component to improve the descent direction of optimal particles, known as the guaranteed convergence particle swarm optimizer (GCPSO) [51], using diversity metrics to ensure swarm diversity [9], and adjusting parameter values to regulate swarm density [54, 55].

In the update equation, if a particle's position becomes close to or overlaps with its pbest or gbest, its motion will depend solely on velocity inertia. When the velocity inertia is zero or near zero, all particles will stop moving, resulting in premature convergence. For instance, simulations on a 10-dimensional Rastrigin function (see Appendix) demonstrate this behavior. Swarm diversity, measured by the variance of the position vectors, is defined by the following equation:

In the update Eq. (2.1), if a particle's position $x_{id}^{(k)}$ becomes close to or overlaps with its pbest $p_{id}^{(k)}$ or gbest $p_{gd}^{(k)}$, its motion will depend solely on velocity inertia. When the velocity inertia is zero or near zero, all particles will stop moving around $p_{gd}^{(k)}$, resulting in premature convergence. For instance, simulations on a 10-dimensional Rastrigin function (shown in Fig. 2.5) demonstrate this behavior. Swarm diversity, measured by the variance of the position vectors, is defined by Eq. (2.37):

$$\bar{x} = \frac{1}{m \cdot D} \sum_{i=1}^{m} \sum_{j=1}^{D} x_{ij}$$

$$Div = \frac{\sum_{i=1}^{m} \sum_{j=1}^{D} [x_{ij} - \bar{x}]^2}{m \cdot D \cdot |ub - lb|} \tag{2.37}$$

where m is the swarm size, D is the dimensionality of the particles, ub and lb are the upper and lower bounds of the search space, respectively.

As shown in Fig. 2.5, during the search within the solution space, when the diversity measure of the swarm stays within a larger range, the fitness value steadily improves. However, once diversity drops to a certain threshold, where particles become highly concentrated or even overlap, further improvement in fitness becomes difficult. Figure 2.6 shows the distribution of particles on the fitness surface. As diversity decreases throughout the optimization, the particles cluster more closely, reducing the area of the search space that the swarm can cover.

This behavior is related to the topology of information transmission among particles. Kennedy [52] investigated and summarized the topology of PSO. In standard

Fig. 2.5 Fitness and diversity in the optimization of a 10-dimensional Rastrigin function using standard PSO

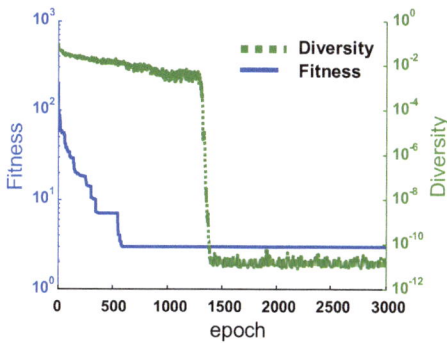

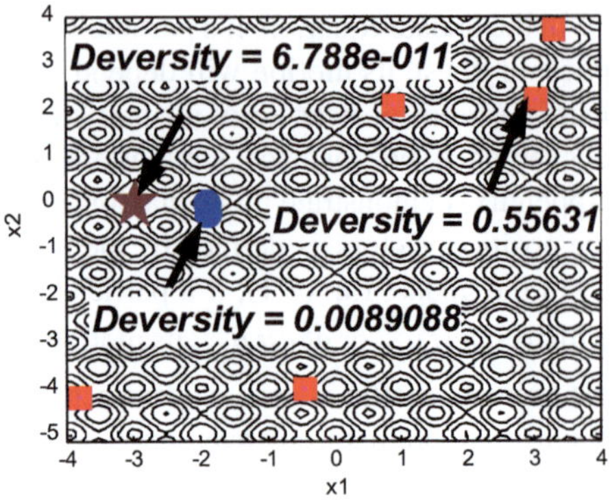

Fig. 2.6 Distribution of five particles with different fitness values in the 2D components of a 10-dimensional fitness surface

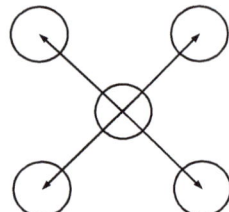

Fig. 2.7 Topology of information transmission in standard PSO

PSO, particles form a typical star-shaped structure by sharing the position information of the gbest particle, as shown in Fig. 2.7. The center represents $gbest\ p_{gd}^{(k)}$, surrounded by four neighbors $x_{id}^{(k)}$. The solid arrows in the figure indicate the direction of information transmission, with Euclidean distance used as the measure of spatial distance between particles. The length of the solid lines represents the distance between particles. The collaboration model between particles follows a "blackboard" approach, emphasizing the processing of a shared optimum while neglecting the exchange of information between individuals. As shown in Figs. 2.5 and 2.6, the distance between particles gradually decreases, and the particles quickly converge around the *gbest* particle. This is one of the key reasons for premature convergence, which leads to the failure of the algorithm.

2.8   Summary

This chapter provides fundamental information about the PSO algorithm, including its description, mathematical models, topological structures, and evaluation metrics. It begins with a formal description of PSO based on the general framework of stochastic search algorithms. Then the chapter summarizes and analyzes various mathematical models proposed by scholars to enhance PSO. Additionally, it explores the topological structures of PSO and concludes by presenting evaluation metrics of

PSO, with a focus on how particle diversity affects the algorithm's performance. This overview sets the stage for the theoretical analysis of PSO in subsequent chapters.

References

1. He XQ, LM, Zhang HG, et al (2000) Sufficient conditions for global stability of a class of fuzzy systems [j]. Control Decis 6:731–733
2. Kennedy J, Eberhart R (1995) Particle swarm optimization. In: Aust F (ed.) Proceedings of the 1995 IEEE International conference on neural networks part 1 (of 6), November 27, 1995–December 1, 1995. IEEE, Perth
3. Kennedy J, Eberhart RC (1997) A discrete binary version of the particle swarm algorithm. In: Proceedings of the 1997 IEEE International conference on systems, man, and cybernetics computational cybernetics and simulation. IEEE, New York, NY, USA
4. Back T, Schwefel HP (1993) An overview of evolution algorithms for parameter optimization [J]. Evol Comput 1
5. Wang ZZ, Bo T (2000) Evolutionary computation [M]. National University of Defense Technology
6. Solis F, Wets R (1981) Minimization by random search techniques [J]. Math Operat Res 6(19–30)
7. Shi Y, Eberhart RC (1998) Parameter selection in particle swarm optimization [M]. 591–600
8. Clerc M, Kennedy J (2002) The particle swarm—explosion, stability, and convergence in a multidimensional complex space [J]. IEEE Trans Evol Comput 6(1):58–73
9. Riget J, Vesterstroem JS (2022) A diversity-guided particle swarm optimizer-the ARPSO [J]. Technical Report No 2002–02. Department of Computer Science, University of Aarhus, EVALife
10. Shi Y, Eberhart R (1998) A modified particle swarm optimizer. In: Proceedings of the 1998 IEEE International conference on evolutionary computation, Proceedings IEEE world congress on computational intelligence. IEEE, New York, NY, USA
11. Clerc M (1999) The swarm and the queen: towards a deterministic and adaptive particle swarm optimization. In: Proceedings of the 1999 congress on evolutionary computation-CEC99. IEEE, Piscataway, NJ, USA
12. Eberhart RC, Shi Y (2000) Comparing inertia weights and constriction factors in particle swarm optimization. In: Proceedings of the 2000 congress on evolutionary computation
13. Kennedy J (2003) Bare bones particle swarms. In: Proceedings of the 2003 IEEE swarm intelligence symposium, 24–26 April 2003. IEEE, Piscataway, NJ, USA
14. Poli R, Langdon WB (2007) Markov chain models of bare-bones particle swarm optimizers. In: Proceedings of the 9th Annual Genetic and Evolutionary Computation Conference, GECCO 2007, July 7, 2007 - July 11, 2007, London, United kingdom, F, 2007 [C]. Association for Computing Machinery.
15. Monson CK, Seppi KD (2005b) Exposing origin-seeking bias in PSO. In: Proceedings of the conference on genetic and evolutionary computation GECCO 2005, June 25, 2005–June 29, 2005. Washington DC, United states. Association for Computing Machinery
16. Feng P, Xiaohui H, Eberhart R, et al. An analysis of bare bones particle swarm. In: Proceedings of the 2008 IEEE swarm intelligence symposium. IEEE, Piscataway, NJ, USA
17. Omran MGH, Engelbrecht A, Salman A (2007a) Barebones particle swarm for integer programming problems. In: Proceedings of the 2007 IEEE swarm intelligence symposium. IEEE, Piscataway, NJ, USA
18. Omran M, Al-Sharhan S (2007) Barebones particle swarm methods for unsupervised image classification. In: Proceedings of the 2007 IEEE congress on evolutionary computation. IEEE, Piscataway, NJ, USA

19. Omran MGH, Engelbrecht AP, Salman A (2007b) Differential evolution based particle swarm optimization. In: Proceedings of the 2007 IEEE swarm intelligence symposium. IEEE, Piscataway, NJ, USA
20. Monson CK, Seppi KD (2005a) Bayesian optimization models for particle swarms. In: Proceedings of the GECCO 2005—genetic and evolutionary computation conference on association for computing machinery, Washington DC, United States
21. Kennedy J, Mendes R (2003) Neighborhood topologies in fully-informed and best-of-neighborhood particle swarms. In: Proceedings of the SMCia/03 and 2003 IEEE International workshop on soft computing in industrial applications. IEEE, Piscataway, NJ, USA
22. Mendes R, Kennedy J, Neves J (2003) Watch thy neighbor or how the swarm can learn from its environment. In: Proceedings of the 2003 IEEE Swarm intelligence symposium. IEEE, Piscataway, NJ, USA
23. Mendes R (2004) Population topologies and their influence in particle swarm performance [D]. Minho, Portugal
24. Fernandez-Martinez JL, Garcia-Gonzalo E, Saraswathi S, et al (2011) Particle swarm optimization: a powerful family of stochastic optimizers. Analysis, design and application to inverse modelling. In: Proceedings of the 2nd International conference on swarm intelligence, ICSI 2011, June 12, 2011–June 15, 2011. Springer Verlag, Chongqing, China
25. Mendes R, Kennedy J, Neves J (2003) Watch thy neighbor or how the swarm can learn from its environment. In: Proceedings of the 2003 IEEE on Swarm intelligence symposium, 2003 SIS '03
26. Chen ZY (2009) Research on neighbor topology structure and algorithm improvement of particle swarm optimization [D]. Chongqing University
27. Kennedy J, Mendes R (2002) Population structure and particle swarm performance. In: Proceedings of the 2002 world congress on computational intelligence—WCCI'02, 12–17 May 2002. IEEE, Piscataway, NJ, USA
28. Suganthan PN (1999) Particle swarm optimizer with neighborhood operator. In: Proceedings of the IEEE congress of evolutionary computation, Washington DC, USA
29. Kennedy J (2000) Stereotyping: improving particle swarm performance with cluster analysis. In: Proceedings of the 2000 congress on evolutionary computation. CEC00 (Cat. No. 00TH8512). IEEE, Piscataway, NJ, USA
30. Peram T, Veeramachaneni K, Mohan CK (2003) Fitness-distance-ratio based particle swarm optimization. In: Proceedings of the 2003 IEEE swarm intelligence symposium. IEEE, Piscataway, NJ, USA
31. Suganthan PN (1999) Particle swarm optimiser with neighbourhood operator. In: Proceedings of the 1999 Congress on evolutionary computation
32. Shi Y, Eberhart R (1998) A modified particle swarm optimization. In: Proceedings of the International conference on evolutionary computation, Anchorage, Alaska
33. Mendes R, Kennedy J, Neves J (2004) The fully informed particle swarm: simpler, maybe better [J]. IEEE Trans Evolut Comput 8(3):204–210
34. Higashi N, Iba H (2003) Particle swarm optimization with Gaussian mutation. In: Proceedings of the 2003 IEEE swarm intelligence symposium. SIS'03 (Cat. No. 03EX706)
35. Shi Y, Eberhart RC (2000) Experimental study of particle swarm optimization. In: Proceedings of the SCI, Orlando, FL
36. Eberhart RC, Xiaohui H (1999) Human tremor analysis using particle swarm optimization. In: Proceedings of the congress on evolutionary computation
37. Pan F, Chen J, Gan GM, et al (2006) Model analysis of particle swarm optimizer [J]. Acta Automat Sin 3:368–377
38. Clerc M (2006) Stagnation analysis in particle swarm optimization or what happens when nothing happens [M]. http://clercmauricefreefr/pso/stagnation analysis
39. Ming J, Yupin L, Shiyuan Y (2007) Stagnation analysis in particle swarm optimization. In: Proceedings of the 2007 IEEE swarm intelligence symposium. IEEE, Piscataway, NJ, USA
40. Parsopoulos KE, Papageorgiou EI, Groumpos PP, et al (2004) Evolutionary computation techniques for optimizing fuzzy cognitive maps in radiation therapy systems. In: Proceedings of the

genetic and evolutionary computation—GECCO 2004 genetic and evolutionary computation conference proceedings, Part I, 26–30 June 2004. Springer-Verlag, Berlin, Germany

41. Riget J, Vesterstr MJS (2002) EVALife Technical Report no, 2002
42. Lovbjerg M, Krink T (2002) Extending particle swarm optimisers with self-organized criticality. In: Proceedings of the 2002 world congress on computational intelligence—WCCI'02. IEEE, Piscataway, NJ, USA
43. Krink T, Vesterstrom JS, Riget J (2002) Particle swarm optimisation with spatial particle extension. In: Proceedings of the 2002 world congress on computational intelligence—WCCI'02. IEEE, Piscataway, NJ, USA
44. Al-Kazemi B, Mohan CK (2002) Multi-phase generalization of the particle swarm optimization algorithm. In: Proceedings of 2002 world congress on computational intelligence–WCCI'02. IEEE, Piscataway, NJ, USA
45. Shi Y, Eberhart RC (2008) Population diversity of particle swarms. In: Proceedings of the 2008 IEEE congress on evolutionary computation (IEEE world congress on computational intelligence), Hong Kong, China
46. Matsui K (1999) New selection method to improve the population diversity in genetic algorithms. In: Proceedings of the IEEE SMC'99 conference on systems, man, and cybernetics. IEEE, Piscataway, NJ, USA
47. Kejun W (2001) A new fuzzy genetic algorithm based on population diversity. In: Proceedings of 2001 International symposium on computational intelligence in robotics and automation. IEEE, Piscataway, NJ, USA
48. Jinyu W, Shaorong W, Shijie C, et al (1998) Measurement based power system load modeling using a population diversity genetic algorithm. In: Proceedings POWERCON '98 1998 International conference on power system technology
49. Wen JY, Wu QH, Shimmin DW, et al (1999) Population diversity based genetic algorithm for fuzzy control of synchronous generators. In: Proceedings of the 1999 IEEE International symposium on computer aided control system design
50. Zheng Z, Yuhui S (2011) Inertia weight adaption in particle swarm optimization algorithm. In: Proceedings of the advances in swarm intelligence second international conference, ICSI 2011. Springer Verlag, Berlin, Germany
51. van den Bergh F (2002) An analysis of particle swarm optimizers [D]. University of Pretoria, South Africa
52. Kennedy J (1999) Small worlds and mega-minds: effects of neighborhood topology on particle swarm performance. In: Proceedings of the 1999 congress on evolutionary computation-CEC99. IEEE, Piscataway, NJ, USA
53. Lovbjerg M, Rasmussen TK, Krink T (2001) Hybrid particle swarm optimiser with breeding and subpopulations. In: Proceedings of the the the third genetic and evolutionary computation conference
54. Parsopoulos KE, Vrahatis MN (2002) Particle swam optimization method in multi-objective problems [J]. ACMS Sympos Appl Comput 603–607
55. Pan F, Tu XY, Chen J, et al (2005) A harmonious particle swarm optimizer——HPSO. Comput Eng 1:169–171

Chapter 3
Algorithm Characteristics of PSO

This chapter delves into the Particle Swarm Optimization (PSO) algorithm by classifying particles into three distinct dynamic models based on their unique update equations and intrinsic properties. We establish a mathematical framework to describe the maximum search space accessible to each particle. Furthermore, by representing the general motion of particles as a weighted sum of historical states, we reveal the inherent "forgetting" behavior of PSO parameters in a probabilistic sense. From the perspective of information transmission, PSO's search strategy can be understood as a probabilistically decaying weighted sum of historical information. The analysis demonstrates that, after a certain period of searching, the standard PSO method becomes probabilistically equivalent to the Bare Bones PSO (BBPS) method.

The three proposed models categorize particles as follows:

(1) **Gbest Model**: Particles that hold the distinction of being both the global best and their own individual best.
(2) **Pbest Model**: Particles that are not the global best but retain their individual best status.
(3) **Common Model**: All remaining particles that are neither the global best nor their individual best.

The classic mathematical model of the PSO algorithm is represented by Eq. (3.1), which can be simplified as follows:

$$v_i^{(k+1)} = w \cdot v_i^{(k)} - \varphi^{(k+1)} \cdot x_i^{(k)} + \varphi^{(k+1)} \cdot p \tag{3.1}$$

$$x_i^{(k+1)} = x_i^{(k)} + \eta \cdot v_i^{(k+1)} \tag{3.2}$$

© Beijing Institute of Technology Press 2025
F. Pan et al., *Particle Swarm Optimizer and Multi-Objective Optimization*,
https://doi.org/10.1007/978-981-95-3381-7_3

where $\varphi^{(k+1)} = \varphi_1^{(k+1)} + \varphi_2^{(k+1)}$ is the acceleration factor, $\varphi_i^{(k+1)} = c_i \cdot r_i$, $p = \frac{\varphi_1^{(k+1)} \cdot p_i + \varphi_2^{(k+1)} \cdot p_l}{\varphi_1^{(k+1)} + \varphi_2^{(k+1)}}$. Furthermore, if we consider p as an external influence on the system, the equation can be rewritten as:

$$x_i^{(k+1)} = \left(1 + \eta \cdot w - \eta \cdot \varphi^{(k+1)}\right) \cdot x_i^{(k)} - \eta \cdot w \cdot x_i^{(k-1)} + \eta \cdot \varphi^{(k+1)} \cdot p \qquad (3.3)$$

3.1 Analysis of the Gbest Model

The analysis of the PSO model encompasses two key aspects: the model's dynamic equation and the concept of the maximum search space with single information. We begin by establishing some crucial definitions:

Definition 3.1 Maximum Search Space with Single Information: This is defined as the farthest space a particle can reach—starting from its current motion state, under the current global and individual best information— assuming no new information is acquired. It is denoted as σ_i.

Definition 3.2 Gbest model: In this model, a PSO particle at a given time step k, is both the global best particle $p_{ld}^{(k)}$ and its individual best particle $p_{id}^{(k)}$, then $x_{id}^{(k)} = p_{id}^{(k)} = p_{ld}^{(k)}$.

Based on **Definition** 3.1 and Eq. (2.1), we derive the dynamic equations for the Gbest model:

$$x_{ld}^{(k+1)} = (1 + w) \cdot x_{ld}^{(k)} - w \cdot x_{ld}^{(k-1)} \qquad (3.4)$$

The characteristic equation and characteristic roots of Eq. (3.4) are given by:

$$D(z) = z^2 - (1 + w) \cdot z + w \qquad z_1 = 1; z_2 = w$$

Assume that at time k_0, the i-th particle becomes the global best particle for the first time, then:

$$\begin{cases} f(x_l^{(k)}) < f(x_l^{(k_l)}) \ \forall k_l < k \\ f(x_l^{(k)}) < f(x_j^{(k)}) \ \forall j \neq l \end{cases}$$

where $f(\cdot)$ is the fitness function, $x_l^{(k)} = (x_{l1}^{(k)}, x_{l2}^{(k)}, \cdots, x_{lN}^{(k)}), v_l^{(k)} = (v_{l1}^{(k)}, v_{l2}^{(k)}, \cdots v_{lN}^{(k)})$, we can derive the following lemma for the maximum search space with single information of Gbest model under the initial conditions $x_l^{(k)}, v_l^{(k)}$.

Lemma 3.1 *The maximum search space with single information of a Gbest model particle is determined by the following equation:*

$$\lim_{n\to\infty} x_{ld}^{(k+n)} = x_{ld}^{(k)} + v_{ld}^{(k)}(1 - \varphi_1^{(k+n)} - \varphi_2^{(k+n)}) \cdot \frac{w^2}{1-w} \tag{3.5}$$

***Proof* Step 1**: First, we provide and prove the iterative description of the Gbest model as shown in Eq. (3.6).

$$x_{ld}^{(k+n)} = x_{ld}^{(k)} + w \cdot v_{ld}^{(k)}\left(w^{n-1} + \left(1 - \varphi_1^{(k+n)} - \varphi_2^{(k+n)}\right) \cdot \sum_{j=1}^{n-2} w^j\right.$$

$$\left. + \prod_{j=2}^{n}\left(1 - \varphi_1^{(k+j)} - \varphi_2^{(k+j)}\right)\right) \tag{3.6}$$

By reorganizing Eqs. (2.1) and (3.4), they can be expressed in the form of Eq. (3.7).

$$\begin{cases} v_{ld}^{(k+1)} = w \cdot v_{ld}^{(k)} \\ x_{ld}^{(k+1)} = x_{ld}^{(k)} + \varphi^{(k+1)} \cdot (p_{ld}^{(k)} - x_{ld}^{(k)}) + v_{ld}^{(k+1)} \end{cases} \tag{3.7}$$

In this case, since the velocity constraint factor η in the standard PSO algorithm typically takes a value of 1, we set it to 1 for simplicity without losing generality. According to Eq. (3.7), we apply mathematical induction to prove Eq. (3.6), while considering the initial condition $x_{id}^{(k)} = p_{id}^{(k)} = p_{ld}^{(k)}$ in Definition 3.2:

(1) **When $n = 1$:**

$$\begin{cases} v_{ld}^{(k+1)} = w \cdot v_{ld}^{(k)} \\ x_{ld}^{(k+1)} = x_{ld}^{(k)} + v_{ld}^{(k+1)} \end{cases}$$

then Eqs. (3.7) and (3.6) holds.

(2) **Assume that Eq. (3.6) holds when $n = m$, then:**

$$x_{ld}^{(k+m)} = x_{ld}^{(k)} + w \cdot v_{ld}^{(k)}\left(w^{m-1} + (1 - \varphi_1^{(k+m)} - \varphi_2^{(k+m)}) \cdot \sum_{j=1}^{m-2} w^j\right.$$

$$\left. + \prod_{j=2}^{m}(1 - \varphi_1^{(k+j)} - \varphi_2^{(k+j)})\right)$$

(3) **Now we prove That Eq. (3.6) also holds when $n = m + 1$:**

$$\begin{cases} v_{ld}^{(k+m+1)} = w^{(m+1)} \cdot v_{ld}^{(k)} \\ x_{ld}^{(k+m+1)} = x_{ld}^{(k+m)} + v_{ld}^{(k+m+1)} + \varphi_1^{(k+m+1)}(p_{ld}^{(m+1)} - x_{ld}^{(k+m)}) \\ \qquad + \varphi_2^{(k+m+1)}(p_{id}^{(m+1)} - x_{ld}^{(k+m)}) \end{cases}$$

$$x_{ld}^{(k+m+1)} = x_{ld}^{(k)} + w \cdot v_{ld}^{(k)} \left(w^{m-1} + \left(1 - \varphi_1^{(k+m)} - \varphi_2^{(k+m)} \right) \cdot \sum_{j=1}^{m-2} w^j \right.$$

$$+ \prod_{j=2}^{m} \left(1 - \varphi_1^{(k+j)} - \varphi_2^{(k+j)} \right) \Bigg) + w^{(m+1)} \cdot v_{ld}^{(k)} - (\varphi_1^{(k+m+1)} + \varphi_2^{(k+m+1)}) \cdot w$$

$$\cdot v_{ld}^{(k)} \left(w^{m-1} + \left(1 - \varphi_1^{(k+m)} - \varphi_2^{(k+m)} \right) \cdot \sum_{j=1}^{m-2} w^j + \prod_{j=2}^{m} \left(1 - \varphi_1^{(k+j)} - \varphi_2^{(k+j)} \right) \right)$$

$$x_{ld}^{(k+m+1)} = x_{ld}^{(k)} + w \cdot v_{ld}^{(k)} \left(w^m + w^{m-1} - \left(\varphi_1^{(k+m+1)} + \varphi_2^{(k+m+1)} \right) \right) \cdot w^{m-1}$$

$$+ \left(1 - \varphi_1^{(k+m)} - \varphi_2^{(k+m)} \right) \cdot \sum_{j=1}^{m-2} w^j - \left(\varphi_1^{(k+m+1)} + \varphi_2^{(k+m+1)} \right) \cdot$$

$$\left(1 - \varphi_1^{(k+m)} - \varphi_2^{(k+m)} \right) \cdot \sum_{j=1}^{m-2} w^j + \prod_{j=2}^{m} \left(1 - \varphi_1^{(k+j)} - \varphi_2^{(k+j)} \right)$$

$$- \left(\varphi_1^{(k+m+1)} + \varphi_2^{(k+m+1)} \right) \cdot \prod_{j=2}^{m} \left(1 - \varphi_1^{(k+j)} - \varphi_2^{(k+j)} \right)$$

$$x_{ld}^{(k+m+1)} = x_{ld}^{(k)} + w \cdot v_{ld}^{(k)} \left(w^m + w^{m-1} \cdot \left(1 - \varphi_1^{(k+m+1)} + \varphi_2^{(k+m+1)} \right) \right.$$

$$+ \left(1 - \varphi_1^{(k+m+1)} + \varphi_2^{(k+m+1)} \right) \cdot \left(1 - \varphi_1^{(k+m)} - \varphi_2^{(k+m)} \right) \cdot \sum_{j=1}^{m-2} w^j$$

$$+ \left(1 - \varphi_1^{(k+m+1)} + \varphi_2^{(k+m+1)} \right) \cdot \prod_{j=2}^{m} \left(1 - \varphi_1^{(k+j)} - \varphi_2^{(k+j)} \right) \Bigg)$$

$$x_{ld}^{(k+m+1)} = x_{ld}^{(k)} + w \cdot v_{ld}^{(k)} \left(w^m + (1 - \varphi_1^{(k+m+1)} + \varphi_2^{(k+m+1)}) \cdot \sum_{j=1}^{m-1} w^j \right.$$

$$+ \prod_{j=2}^{m+1} (1 - \varphi_1^{(k+j)} - \varphi_2^{(k+j)}) \Bigg)$$

As shown in the above equation, Eq. (3.6) holds.

Step 2: Based on Eq. (3.6), we derive the conclusion of Eq. (3.5).

For convenience, let $\chi = w^{n-1} + (1 - \varphi_1^{(k+n)} - \varphi_2^{(k+n)}) \cdot \sum_{j=1}^{n-2} w^j +$

$\prod_{j=2}^{n} (1 - \varphi_1^{(k+j)} - \varphi_2^{(k+j)}).$

When $n \to \infty$, since $0 < w < 1$, the first term on the right side of χ can be expressed as follows:

$$\lim_{n \to \infty} w^{n-1} = 0$$

Since the second term of χ is a geometric series, we have:

$$\lim_{n \to \infty} (1 - \varphi_1^{(k+n)} - \varphi_2^{(k+n)}) \cdot \sum_{j=1}^{n-2} w^j = (1 - \varphi_1^{(k+n)} - \varphi_2^{(k+n)}) \cdot \frac{w}{1-w}$$

The limit of the third term on the right side of χ is given by:

$$\lim_{n \to \infty} \prod_{j=2}^{n+1} (1 - \varphi_1^{(k+j)} - \varphi_2^{(k+j)}) = 0$$

The result of the above equation is 0 because $\varphi_i^{(k+j)}$ is uniformly distributed between $(0, 1)$, so $(1 - \varphi_1^{(k+j)} - \varphi_2^{(k+j)})$ is uniformly distributed between $(-3, 1)$. Consequently, when $n \to \infty$, then $\lim_{n \to \infty} P((1 - \varphi_1^{(k+j)} - \varphi_2^{(k+j)}) \neq 0) = 0$, which means that there will always be a moment when $(1 - \varphi_1^{(k+j)} - \varphi_2^{(k+j)})$ takes on a value of 0, and the corresponding part of the factorial term will be 0. Combining the above three analyses, we can archive the final result as follows:

$$\lim_{n \to \infty} x_{ld}^{(k+n)} = x_{ld}^{(k)} + v_{ld}^{(k)}(1 - \varphi_1^{(k+n)} - \varphi_2^{(k+n)}) \cdot \frac{w^2}{1-w}$$

The maximum search space of *Gbest* is illustrated in Fig. 3.1. Lemma 3.1 indicates that particles in the *Gbest* model, due to containing the optimal information of the swarm, tend to remain within an area centered around the initial optimal position vector $x_{ld}^{(k)}$ during the search process. The size of this area is proportional to the initial velocity vector $v_{ld}^{(k)}$. The probability that this region covers previously searched space is greater than that of exploring new regions. This is because when $(1 - \varphi_1^{(k+j)} - \varphi_2^{(k+j)})$ is negative, it reverses the direction of $v_{ld}^{(k)}$.

From the above analysis, if the swarm's best particle does not obtain new information, it will quickly stagnate at a point within a finite period and lose the capability to further explore other spaces. In the *Gbest* model, two special cases are noted:

(1) When the velocity $v_{ld}^{(k)} = 0$, or the inertia factor $w = 0$, $x_{ld}^{(k+1)} = x_{ld}^{(k_0)}$, the particle loses its ability to move and falls into a stagnated state at a fixed point.
(2) When $w = 1$, the characteristic equation has two repeated eigenvalues $z_1 = z_2 = 1$. In this case, the particle moves in a straight line at a constant speed from the initial position according to its initial velocity.

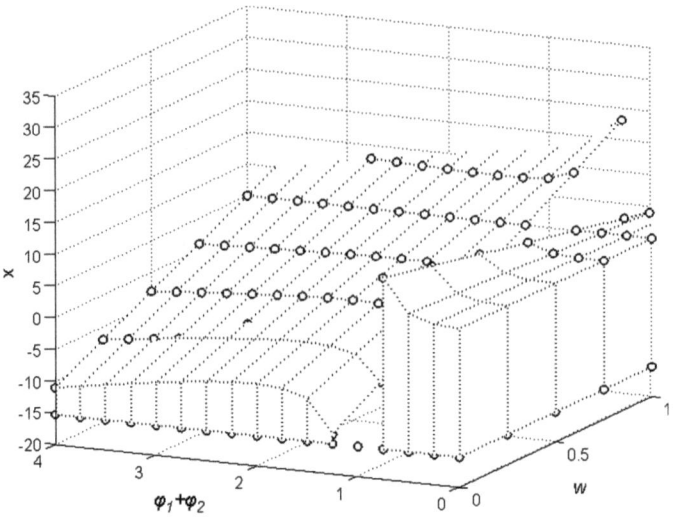

Fig. 3.1 Maximum search space achievable by the Gbest model with single-information conditions

3.2   Analysis of the Pbest Model

Definition 3.3 Pbest model: The *Pbest* model is defined as a PSO particle at time k that is the individual best particle $p_{id}^{(k)}$, but not the global best particle $p_{ld}^{(k)}$, that means $x_{id}^{(k)} = p_{id}^{(k)} \neq p_{ld}^{(k)}$.

According to Definition 3.3, the *Pbest* model describes a scenario where the global best information $p_{ld}^{(k)}$ remains unchanged while the particle continually improves its individual best information. Combining Eq. (2.1), the dynamic equation of the *Pbest* model can be obtained as:

$$x_{id}^{(k+1)} = (1 + w - \varphi_2^{(k+1)}) \cdot x_{id}^{(k)} - w \cdot x_{id}^{(k-1)} + \varphi_2^{(k+1)} \cdot p_{ld}^{(k)} \tag{3.8}$$

The characteristic equation of Eq. (3.8) is:

$$D(z) = z^2 - (1 + w - \varphi_2^{k+1}) \cdot z + w$$

Under the single-information condition, the possible maximum search space of a *Pbest* model particle consists of two parts: one is the hypercube $\varphi_2^{(k+1)}(p_{gd}^{(k)} - x_{id}^{(k)})$ formed by the global best particle and the *Pbest* particle in the search space, and the other is the search space synthesized with the particle's velocity vector $v_i^{(k)}$. Figure 3.2 shows the possible maximum search space of a *Pbest* model under the single-information condition in a two-dimensional search space.

The initial conditions are $x_i^{(k)} = (3, 5), p_{ld}^{(k)} = (4, 1), v_i^{(k)} = (-1, -3), w \in [0, 1]$, $\varphi_2^{(k)} \in [0, 2]$. "△" stands for the spatial position of the global best particle $p_{ld}^{(k)}$, and

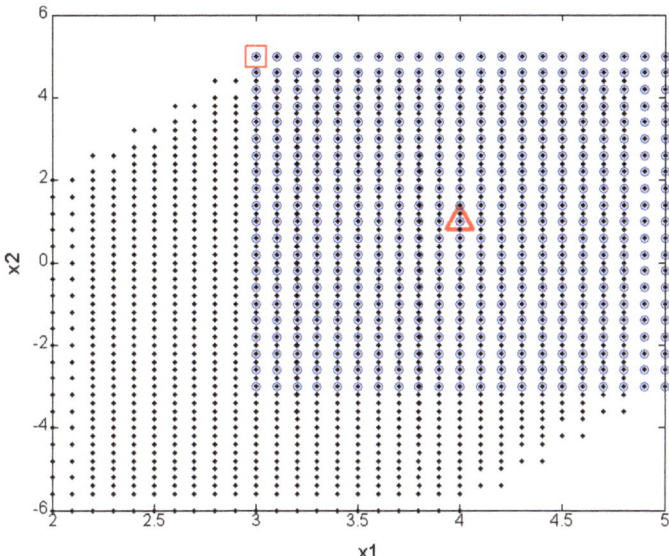

Fig. 3.2 Maximum search space achievable by the Pbest model with single-information conditions

"□" indicates the spatial position of the *Pbest* model particle $x_i^{(k)}$. The hypercube formed by "⊙" represents the search space covered by $\varphi_2^{(k+1)}(p_{ld}^{(k)} - x_{id}^{(k)})$, and the polyhedron formed by "·" is the maximum possible search space that can be covered by the synthesis of position vector $\varphi_2^{(k+1)}(p_{ld}^{(k)} - x_{id}^{(k)})$ and the velocity vector $v_i^{(k)}$.

Lemma 3.2 *The maximum search space with single information of a Pbest model particle is determined by the following equation:*

$$
\lim_{n \to \infty} x_{id}^{(k+n)} = w \cdot v_{id}^{(k)} \cdot \left(w^{n-1} + \sum_{j=2}^{n} w^{n-j} \cdot \prod_{l=1}^{j} (1 - \varphi_2^{(k+n-l+1)}) \right)
$$

$$
+ p_{ld}^{(k)} \cdot \left(\varphi_2^n + \sum_{j=2}^{n} \varphi_2^{(k+n-j+1)} \prod_{l=1}^{j-1} (1 - \varphi_2^{(n-l+1)}) \right) \tag{3.9}
$$

Proof According to Eq. (3.9), the iterative description of the *Pbest* model can be written as:

$$x_{id}^{(k+n)} = \prod_{j=1}^{n}(1 - \varphi_2^j) \cdot x_{id}^{(k)} + w \cdot v_{id}^{(k)} \cdot (w^{n-1} + \sum_{j=2}^{n} w^{n-j} \cdot \prod_{l=1}^{j}(1 - \varphi_2^{(k+n-l+1)}))$$

$$+ p_{ld}^{(k)} \cdot (\varphi_2^{(k+n)} + \sum_{j=2}^{n} \varphi_2^{(n-j+1)} \prod_{l=1}^{j-1}(1 - \varphi_2^{(k+n-l+1)}))$$

(3.10)

For convenience in proof and expression, we rewrite Eq. (3.8) as follows:

$$\begin{cases} v_{id}^{(k+1)} = w \cdot v_{id}^{(k)} \\ x_{id}^{(k+1)} = x_{id}^{(k)} + \varphi_2^{(k+1)} \cdot (p_{ld}^{(k)} - x_{id}^{(k)}) + v_{id}^{(k+1)} \end{cases}$$

(3.11)

Since the velocity proportion constraint factor η in the standard PSO algorithm is typically set to 1, we also set it to 1 for simplicity without loss of generality. Based on Eq. (3.11), we use mathematical induction to prove Eq. (3.10):

(1) **When $n= 1$,**

$$x_{id}^{(k+1)} = (1 - \varphi_2^{(k+1)}) \cdot x_{id}^{(k)} + w \cdot v_{id}^{(k)} + \varphi_2^{(k+1)} \cdot p_{ld}^{(k)}$$

Equation (3.10) holds.

(2) **Assume that Eq. (3.10) holds when $n= m$, we have:**

$$x_{id}^{(k+m)} = \prod_{j=1}^{m}(1 - \varphi_2^{(k+j)}) \cdot x_{id}^{(k)} + w \cdot v_{id}^{(k)} \cdot (w^{m-1} + \sum_{j=2}^{m} w^{m-j} \cdot \prod_{l=1}^{j}(1 - \varphi_2^{(k+m-l+1)}))$$

$$+ p_{ld}^{(k)} \cdot (\varphi_2^{(k+m)} + \sum_{j=2}^{m} \varphi_2^{(m-j+1)} \prod_{l=1}^{j-1}(1 - \varphi_2^{(k+m-l+1)}))$$

(3) **then, when $n= m + 1$, we have:**

$$x_{id}^{(k+m+1)} = x_{id}^{(k+m)} + \varphi_2^{(k+m+1)} \cdot (p_{ld}^{(k)} - x_{id}^{(k+m)}) + v_{id}^{(k+m+1)}$$

$$x_{id}^{(k+m+1)} = (1 - \varphi_2^{(k+m+1)})x_{id}^{(k+m)} + \varphi_2^{(k+m+1)} p_{ld}^{(k)} + v_{id}^{(k+m+1)}$$

$$x_{id}^{(k+m+1)} = (1 - \varphi_2^{(k+m+1)}) \cdot \left\{ \prod_{j=1}^{m}(1 - \varphi_2^{(k+j)}) \cdot x_{id}^{(k)} + w \cdot v_{id}^{(k)} \cdot \left(w^{m-1} + \right. \right.$$

$$\left. \sum_{j=2}^{m} w^{n-j} \cdot \prod_{l=1}^{j}\left(1 - \varphi_2^{(k+m-l+1)}\right)\right) + p_{ld}^{(k)} \cdot \left(\varphi_2^{(k+m)} + \sum_{j=2}^{m} \varphi_2^{(k+m-j+1)}\right.$$

$$\prod_{l=1}^{j-1}(1 - \varphi_2^{(k+m-l+1)})\bigg)\bigg\} + \varphi_2^{(k+m+1)}p_{ld}^{(k)} + w^{m+1}v_{id}^{(k)}$$

$$x_{id}^{(k+m+1)} = \left(1 - \varphi_2^{(k+m+1)}\right) \cdot \prod_{j=1}^{m}\left(1 - \varphi_2^{(k+j)}\right) \cdot x_{id}^{(k)}$$

$$+ w \cdot v_{id}^{(k)} \cdot \left(w^m + \left(1 - \varphi_2^{(k+m+1)}\right) \cdot w^{m-1} + \sum_{j=2}^{m} w^{m-j} \cdot \prod_{l=1}^{j}(1 - \varphi_2^{(k+m-l+1)})\right)$$

$$+ p_{ld}^{(k)} \cdot \left(\varphi_2^{(k+m+1)} + \left(1 - \varphi_2^{(k+m+1)}\right)\right) \cdot \varphi_2^{(k+m)} + \left(1 - \varphi_2^{(k+m+1)}\right) \cdot \sum_{j=2}^{m} \varphi_2^{(k+m-j+1)}$$

$$\prod_{l=1}^{j-1}(1 - \varphi_2^{(k+m-l+1)})\bigg)$$

$$x_{id}^{(k+m+1)} = \prod_{j=1}^{m+1}(1 - \varphi_2^{(k+j)}) \cdot x_{id}^{(k)} + w \cdot v_{id}^{(k)} \cdot (w^m + \sum_{j=2}^{m+1} w^{m-j} \cdot \prod_{l=1}^{j}(1 - \varphi_2^{(k+m-l+1)})$$

$$+ p_{ld}^{(k)} \cdot (\varphi_2^{(m+1)} + \sum_{j=2}^{m+1}\varphi_2^{(k+m-j+1)}\prod_{l=1}^{j-1}(1 - \varphi_2^{(k+m-l+1)}))$$

This is consistent with Eq. (3.10) when $n = m + 1$. Therefore, Eq. (3.10) is proven. Further, considering when $n \to \infty$ or $n >> k$ is present, and the range of $\varphi_2^{(k+1)}$ in the standard PSO algorithm is $\varphi_2^{(k+1)} \in (0, 2)$, then $(1 - \varphi_2^{(k+j)}) \in (-1, 1)$. The right-hand side of Eq. (3.10) is:

$$\lim_{m \to \infty}\left(\prod_{j=1}^{n}\left(1 - \varphi_2^{(k+j)}\right) \cdot x_{id}^{(k)}\right) = 0$$

Thus, the maximum search space with single information of a *Pbest* model particle is independent of the initial position vector, but is instead determined by the last two terms on the right-hand side of Eq. (3.10).

Now consider the term $v_{id}^{(k)}$ on the right-hand side of Eq. (3.10):

$$w \cdot \left(w^{n-1} + \sum_{j=2}^{n} w^{n-j} \cdot \prod_{l=1}^{j}\left(1 - \varphi_2^{(k+n-l+1)}\right)\right) << \frac{w}{1 - w} \tag{3.12}$$

A qualitative analysis of v and w can be performed using Eq. (3.9): When $w = 0.5$, $v_{id}^{(k)}$ influences the final $x_{id}^{(k)}$ with its original velocity value. When $0 < w < 0.5$, Eq. (3.12) is less than 1, reducing the influence of $v_{id}^{(k)}$ on $x_{id}^{(k)}$ and strengthening the particle's convergence trend towards $p_{ld}^{(k)}$. This convergence trend becomes more pronounced as w decreases.

When $0.5 < w < 1$, Eq. (3.12) is greater than 1, strengthening the influence of $v_{id}^{(k)}$ on x_{id}^{k} and weakening the particle's convergence trend towards $p_{ld}^{(k)}$. The larger the value of w, the further the particle is from $p_{ld}^{(k)}$.

$w = 0$ and $w = 1$ are two special cases: one completely ignores $v_{id}^{(k)}$, while the other causes $v_{id}^{(k)}$ to have no decay so that the particle will continue to move indefinitely.

Riget suggests that w linearly decreases with the increase of iteration cycles during the optimization process. And the above analysis provides a reasonable explanation for this.

Furthermore, $\varphi_2^{(k+1)}$ is a uniformly distributed random number between $(0, 2)$. For the $p_{ld}^{(k)}$ coefficient term, it causes the maximum search space with single information of $x_{id}^{(k)}$ probabilistically distributed around $p_{ld}^{(k)}$.

3.3 Description of the Standard PSO Maximum Search Space with Single Information

Sections 3.1 and 3.2 derive the maximum search space under single information condition for the two special cases (Gbest and Pbest particle models) without the introduction of new information. The results explain that typical parameter settings cause the swarm to search already explored regions repeatedly. This behavior reduces search efficiency and incurs additional computational costs.

Lemma 3.3 *Provides the general movement characteristics and the maximum achievable search space for arbitrary particle in standard PSO (in the Common model: a particle that is neither the gbest nor the pbest) at any given time.*

Lemma 3.4 *The maximum search space with single information of a standard PSO particle is given by the following formula:*

$$
\begin{aligned}
x_{id}^{(k+n)} = {} & \left\{ \varphi_2^{(k+n)} + \sum_{j=1}^{n-1} \left(\varphi_2^{(k+j)} \cdot \prod_{m=j+1}^{n} \gamma(m) \right) \right\} \cdot p_{ld}^{(k)} \\
& + \left\{ \varphi_1^{(k+n)} + \sum_{j=1}^{n-1} \left(\varphi_1^{(k+j)} \cdot \prod_{m=j+1}^{n} \gamma(m) \right) \right\} \cdot p_{id}^{(k)} + \left\{ \prod_{m=1}^{n} \gamma(m) \right\} \cdot x_{id}^{(k)} \\
& + \left\{ w^n + \sum_{j=1}^{n-1} \left(w^j \cdot \prod_{m=j+1}^{n} \gamma(m) \right) \right\} v_{id}^{(k)} +
\end{aligned}
\tag{3.13}
$$

where $\gamma(m) = 1 - \varphi_1^{(k+m)} - \varphi_2^{(k+m)}$, $m = 1, 2, \cdots, n$.

Note : The superscript of a variable with parentheses denotes the iteration cycle, such as $x_{id}^{(k+n)}$ represents the $k+n$ cycle; the superscript of a variable without parentheses denotes the exponents, such as w^n indicates w raised to the power of n.

Proof Combining Eqs. (3.1) and (3.2), we get the following equation:

$$v_{id}^{(k+1)} = w \cdot v_{id}^{(k)} \tag{3.14}$$

$$x_{id}^{(k+1)} = x_{id}^{(k)} + \varphi_2^{(k+1)} \cdot (p_{ld}^{(k)} - x_{id}^{(k)}) + \varphi_1^{(k+1)} \cdot (p_{id}^{(k)} - x_{id}^{(k)}) + v_{id}^{(k+1)} \tag{3.15}$$

when $n = 1$, $x_{id}^{(k+1)} = \gamma(1) \cdot x_{id}^{(k)} + \varphi_2^{(k+1)} \cdot p_{ld}^{(k)} + \varphi_1^{(k+1)} \cdot p_{id}^{(k)} + w \cdot v_{id}^{(k)}$, where $\gamma(m) = 1 - \varphi_1^{(k+m)} - \varphi_2^{(k+m)}$., $m = 1, 2, \cdots, n$.

when $n = 2$, we have:

$$x_{id}^{(k+2)} = \gamma(2) \cdot x_{id}^{(k+1)} + \varphi_2^{(k+2)} \cdot p_{ld}^{(k)} + \varphi_1^{(k+2)} \cdot p_{id}^{(k)} + w^2 \cdot v_{id}^{(k)}$$

$$x_{id}^{(k+2)} = \gamma(2) \cdot \varphi_2^{(k+1)} \cdot p_{ld}^{(k)} + \varphi_2^{(k+2)} \cdot p_{ld}^{(k)} + \gamma(2) \cdot \varphi_1^{(k+1)} \cdot p_{id}^{(k)} + \varphi_1^{(k+2)} \cdot p_{id}^{(k)}$$
$$+ \gamma(2) \cdot \gamma(1) \cdot x_{id}^{(k)} + \gamma(2) \cdot w \cdot v_{id}^{(k)} + w^2 \cdot v_{id}^{(k)}$$

$$x_{id}^{(k+2)} = \left\{ \varphi_2^{(k+2)} + \sum_{j=1}^{1} \left(\varphi_2^{(k+j)} \cdot \prod_{m=j+1}^{2} \gamma(m) \right) \right\} \cdot p_{ld}^{(k)} + \left\{ \varphi_1^{(k+2)} + \sum_{j=1}^{1} \left(\varphi_1^{(k+j)} \cdot \right. \right.$$

$$\left. \left. \prod_{m=j+1}^{2} \gamma(m) \right) \right\} \cdot p_{id}^{(k)} + \left\{ \prod_{m=1}^{2} \gamma(m) \right\} \cdot x_{id}^{(k)} + \left\{ w^2 + \sum_{j=1}^{1} \left(w^j \cdot \prod_{m=j+1}^{n} \gamma(m) \right) \right\} v_{id}^{(k)}$$

Assuming that Eq. (3.14) holds for $n = m$:

$$x_{id}^{(k+m)} = \left\{ \varphi_2^{(k+m)} + \sum_{j=1}^{m-1} \left(\varphi_2^{(k+j)} \cdot \prod_{m=j+1}^{m} \gamma(m) \right) \right\} \cdot p_{ld}^{(k)} + \left\{ \varphi_1^{(k+m)} + \sum_{j=1}^{m-1} \left(\varphi_1^{(k+j)} \right. \right.$$

$$\left. \left. \cdot \prod_{m=j+1}^{m} \gamma(m) \right) \right\} \cdot p_{id}^{(k)} + \left\{ \prod_{m=1}^{m} \gamma(m) \right\} \cdot x_{id}^{(k)} + \left\{ w^m + \sum_{j=1}^{m-1} \left(w^j \cdot \prod_{m=j+1}^{m} \gamma(m) \right) \right\} v_{id}^{(k)}$$

Assuming that for $n = m + 1$, we have

$$x_{id}^{(k+m+1)} = x_{id}^{(k+m)} + \varphi_2^{(k+m+1)} \cdot (p_{ld}^{(k)} - x_{id}^{(k+m)}) + \varphi_1^{(k+m+1)} \cdot (p_{id}^{(k)} - x_{id}^{(k+m)}) + v_{id}^{(k+m+1)}$$
$$= \gamma(m+1) \cdot x_{id}^{(k+m)} + \varphi_2^{(k+m+1)} \cdot p_{ld}^{(k)} + \varphi_1^{(k+m+1)} \cdot p_{id}^{(k)} + w^{m+1} \cdot v_{id}^{(k)}$$

$$
\begin{aligned}
x_{id}^{(k+m+1)} &= \gamma(m+1) \cdot \left\{ \varphi_2^{(k+m)} + \sum_{j=1}^{m-1} \left(\varphi_2^{(k+j)} \cdot \prod_{m=j+1}^{m} \gamma(m) \right) \right\} \cdot p_{ld}^{(k)} + \varphi_2^{(k+m+1)} \cdot p_{ld}^{(k)} \\
&+ \gamma(m+1) \cdot \left\{ \varphi_1^{(k+m)} + \sum_{j=1}^{m-1} \left(\varphi_1^{(k+j)} \cdot \prod_{m=j+1}^{m} \gamma(m) \right) \right\} \cdot p_{id}^{(k)} + \varphi_1^{(k+m+1)} \cdot p_{id}^{(k)} \\
&+ \gamma(m+1) \cdot \left\{ \prod_{m=1}^{m} \gamma(m) \right\} \cdot x_{id}^{(k)} + \gamma(m+1) \cdot \left\{ w^m + \sum_{j=1}^{m-1} \left(w^j \cdot \prod_{m=j+1}^{m} \gamma(m) \right) \right\} \\
& v_{id}^{(k)} + w^{m+1} \cdot v_{id}^{(k)}
\end{aligned}
$$

$$
\begin{aligned}
x_{id}^{(k+m+1)} &= \left\{ \varphi_2^{(k+m+1)} + \sum_{j=1}^{m} \left(\varphi_2^{(k+j)} \cdot \prod_{m=j+1}^{m+1} \gamma(m) \right) \right\} \cdot p_{ld}^{(k)} + \left\{ \varphi_1^{(k+m+1)} + \sum_{j=1}^{m} \left(\varphi_1^{(k+j)} \cdot \right. \right. \\
& \left. \prod_{m=j+1}^{m+1} \gamma(m) \right) \right\} \cdot p_{id}^{(k)} + \left\{ \prod_{m=1}^{m+1} \gamma(m) \right\} \cdot x_{id}^{(k)} + \left\{ w^{m+1} + \sum_{j=1}^{m} \left(w^j \cdot \prod_{m=j+1}^{m+1} \gamma(m) \right) \right\} v_{id}^{(k)}
\end{aligned}
$$

Thus, Eq. (3.13) is proven.

3.4   Similarity Analysis Between Standard PSO and BBPS

3.4.1   Description and Analysis of Maximum Search Space with Single Information

We analyze and derive the Eq. (3.13). For clarity, we first consider the auxiliary variable $\hat{x}_{id}^{(k+n)}$ that is related only to parameters $p_{ld}^{(k)}$ and $p_{id}^{(k)}$ in the following derivation:

$$
\begin{aligned}
\hat{x}_{id}^{(k+n)} &= \left\{ \varphi_{ld}^{(k+n)} + \sum_{j=1}^{n-1} \left(\varphi_{ld}^{(k+j)} \cdot \prod_{m=j+1}^{n} \gamma(m) \right) \right\} \cdot p_{ld}^{(k)} \\
&+ \left\{ \varphi_{id}^{(k+n)} + \sum_{j=1}^{n-1} \left(\varphi_{id}^{(k+j)} \cdot \prod_{m=j+1}^{n} \gamma(m) \right) \right\} \cdot p_{id}^{(k)}
\end{aligned} \tag{3.16}
$$

$$
\begin{aligned}
\hat{x}_{id}^{(k+n)} &= \left\{ \varphi_{ld}^{(k+n)} + \varphi_{id}^{(k+n)} + \sum_{j=1}^{n-1} \left(\prod_{m=j+1}^{n} \gamma(m) \right) - \sum_{j=1}^{n-1} \left(\gamma(j) \cdot \prod_{m=j+1}^{n} \gamma(m) \right) \right\} \cdot p_{ld}^{(k)} \\
&+ \left\{ \varphi_{id}^{(k+n)} + \sum_{j=1}^{n-1} \left(\varphi_{id}^{(k+j)} \cdot \prod_{m=j+1}^{n} \gamma(m) \right) \right\} \cdot (p_{id}^{(k)} - p_{ld}^{(k)})
\end{aligned}
$$

$$
\hat{x}_{id}^{(k+n)} = \left\{ \varphi_{ld}^{(k+n)} + \varphi_{id}^{(k+n)} + \sum_{j=1}^{n-1} \left(\prod_{m=j+1}^{n} \gamma(m) \right) - \sum_{j=1}^{n-1} \left(\prod_{m=j}^{n} \gamma(m) \right) \right\} \cdot p_{ld}^{(k)}
$$

$$+ \left\{ \varphi_{id}^{(k+n)} + \sum_{j=1}^{n-1} \left(\varphi_{id}^{(k+j)} \cdot \prod_{m=j+1}^{n} \gamma(m) \right) \right\} \cdot (p_{id}^{(k)} - p_{ld}^{(k)})$$

The cumulative term of the $p_{ld}^{(k)}$ parameter in the above formula can be expanded by the cycle j in Fig. 3.3. As shown in the figure, the corresponding terms of the two cumulative series cancel each other out in each adjacent interval cycle, ultimately retaining the terms corresponding to $j = 1$ and $j = n-1$.

After the cancellation of terms in Fig. 3.3, $\hat{x}_{id}^{(k+n)}$ can be more directly interpreted as: A random search within a space centered on the global best vector, with a radius equal to the distance between the global optimal vector and the individual historical best vector, which conforms to a certain probability distribution.

Based on the above derivation, a comprehensive expression for all terms in Eq. (3.13) is given as follows:

$$x_{id}^{(k+n)} = p_{ld}^{(k)} - \prod_{m=1}^{n} \gamma(m) \cdot p_{ld}^{(k)} + \left\{ \varphi_{id}^{(k+n)} + \sum_{j=1}^{n-1} \left(\varphi_{id}^{(k+j)} \cdot \prod_{m=j+1}^{n} \gamma(m) \right) \right\}$$
$$\cdot (p_{id}^{(k)} - p_{ld}^{(k)}) + \left\{ \prod_{m=1}^{n} \gamma(m) \right\} \cdot x_{id}^{(k)} + \left\{ w^n + \sum_{j=1}^{n-1} \left(w^j \cdot \prod_{m=j+1}^{n} \gamma(m) \right) \right\} v_{id}^{(k)}$$

This expression can be further simplified as:

$$x_{id}^{(k+n)} = p_{ld}^{(k)} + \phi^{(n)} \cdot \left(p_{id}^{(k)} - p_{ld}^{(k)} \right) - Y_{(n)} \cdot p_{ld}^{(k)} + Y_{(n)} \cdot x_{id}^{(k)} + \delta^{(n)} \cdot v_{id}^{(k)} \quad (3.17)$$

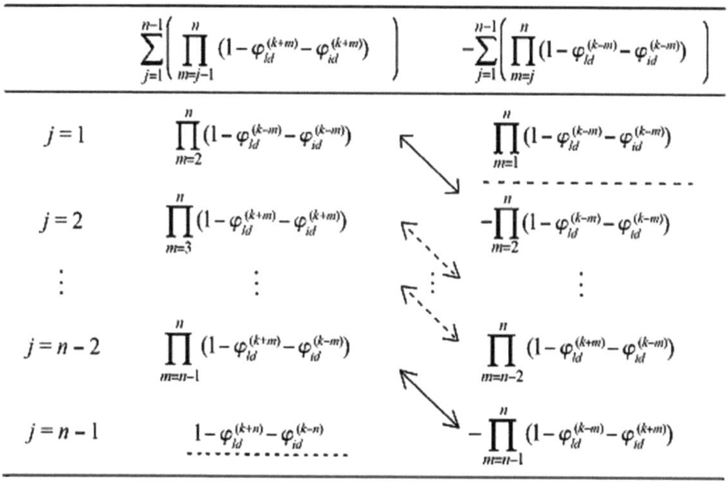

	$\sum_{j=1}^{n-1} \left(\prod_{m=j-1}^{n} (1-\varphi_{id}^{(k+m)} - \varphi_{id}^{(k+m)}) \right)$	$-\sum_{j=1}^{n-1} \left(\prod_{m=j}^{n} (1-\varphi_{id}^{(k-m)} - \varphi_{id}^{(k-m)}) \right)$
$j=1$	$\prod_{m=2}^{n} (1-\varphi_{id}^{(k-m)} - \varphi_{id}^{(k-m)})$	$\prod_{m=1}^{n} (1-\varphi_{id}^{(k-m)} - \varphi_{id}^{(k-m)})$
$j=2$	$\prod_{m=3}^{n} (1-\varphi_{id}^{(k+m)} - \varphi_{id}^{(k+m)})$	$-\prod_{m=2}^{n} (1-\varphi_{id}^{(k-m)} - \varphi_{id}^{(k-m)})$
$\vdots$	$\vdots$	$\vdots$
$j=n-2$	$\prod_{m=n-1}^{n} (1-\varphi_{id}^{(k+m)} - \varphi_{id}^{(k-m)})$	$\prod_{m=n-2}^{n} (1-\varphi_{id}^{(k+m)} - \varphi_{id}^{(k-m)})$
$j=n-1$	$1-\varphi_{id}^{(k+n)} - \varphi_{id}^{(k-n)}$	$-\prod_{m=n-1}^{n} (1-\varphi_{id}^{(k-m)} - \varphi_{id}^{(k+m)})$

Fig. 3.3 Diagram of parameter $p_{ld}^{(k)}$ cumulative term expansion

where: $Y_{(n)} = \prod_{m=1}^{n} \gamma(m)$ is the factorial decay factor of $p_{ld}^{(k)}$ and $x_{id}^{(k)}$; $\phi^{(n)} =$

$\left\{ \varphi_{id}^{(k+n)} + \sum_{j=1}^{n-1} \left(\varphi_{id}^{(k+j)} \cdot \prod_{m=j+1}^{n} \gamma(m) \right) \right\}$ is the random weight of the distance between

$p_{ld}^{(k)}$ and $p_{id}^{(k)}$; $\delta^{(n)} = \left\{ w^n + \sum_{j=1}^{n-1} \left(w^j \cdot \prod_{m=j+1}^{n} \gamma(m) \right) \right\}$ is the weighted parameter of
the initial velocity vector.

We will discuss and analysis these three parameters separately to demonstrate that after a sufficiently large number of iterations n, $x_{id}^{(k+n)}$ becomes equivalent to $\hat{x}_{id}^{(k+n)}$.

3.4.2 Analysis of the Factorial Decay Factor for the Initial Position Vector

The factorial decay factor $Y_{(n)} = \prod_{m=1}^{n} \gamma(m)$ reflects how the initial position information $x_{id}^{(k)}$ and part of the gbest information $p_{ld}^{(k)}$ evolve over time. Without loss of generality, based on the definition in Eq. (1.1), we assume that $\varphi_{ld}^{(k)} \sim (0, r_l)$ and $\varphi_{id}^{(k)} \sim (0, r_i)$ are uniformly distributed, and $\varphi_{ld}^{(k)}$ and $\varphi_{id}^{(k)}$ are statistically independent. Thus, we have.

The factorial decay factor $Y_{(n)} = \prod_{m=1}^{n} \gamma(m)$ reflects how the particle's initial position information $x_{id}^{(k)}$ and partial swarm best information $p_{ld}^{(k)}$ diminish over time. Without loss of generality, according to the definition in Eq. (2.1), $\varphi_{ld}^{(k)} \sim (0, r_l), \varphi_{id}^{(k)} \sim (0, r_i)$ are uniformly distributed, and $\varphi_{ld}^{(k)}$ and $\varphi_{id}^{(k)}$ are independent. Therefore, we have:

$$E[X_{(m)}] = 1 - E\left[\varphi_{ld}^{(k+m)}\right] - E\left[\varphi_{id}^{(k+m)}\right] = 1 - \frac{r_l + r_i}{2}$$

$$D[X_{(m)}] = D\left[\varphi_{ld}^{(k+m)}\right] + D\left[\varphi_{id}^{(k+m)}\right] = \frac{r_l^2 + r_i^2}{12}$$

$$E[X_{(m)}^2] = D[X_{(m)}] + \left(E[X_{(m)}]\right)^2 = 1 + \frac{r_l^2 + r_i^2}{3} + \frac{r_l r_i}{2} - r_l - r_i$$

$X_{(j)}, X_{(j+1)}, \ldots, X_{(n)}$ are statistically independent, and the probability density function can be expressed as:

$$f_{X_{(j)}X_{(j+1)}\cdots X_{(n)}}(x_{(j)}, \cdots, x_{(n)}) = f_{X_{(j)}}(x_{(j)}) \cdot f_{X_{(j+1)}}(x_{(j+1)}) \cdots f_{X_{(n)}}(x_{(n)})$$

$$E[Y_{(n)}] = E\left[\prod_{m=j}^{n} \gamma(m)\right] = E\left[\prod_{m=j}^{n} \gamma(m)\right] = \prod_{m=j}^{n} E[\gamma(m)] = (1 - \frac{r_l + r_i}{2})^{n-j}$$

(3.18)

$$E[Y_{(n)}^2] = E\left[\left(\prod_{m=j}^{n} \gamma(m)\right)^2\right] = E\left[\prod_{m=j}^{n} \gamma(m)^2\right] = \prod_{m=j}^{n} E[\gamma(m)^2]$$

$$= \left(1 + \frac{r_l^2 + r_i^2}{3} + \frac{r_l r_i}{2} - r_l - r_i\right)^{n-j} \quad D[Y_{(n)}] = E[Y_{(n)}^2] - (E[Y_{(n)}])^2$$

$$= \left(1 + \frac{r_l^2 + r_i^2}{3} + \frac{r_l r_i}{2} - r_l - r_i\right)^{n-j} - \left(1 - \frac{r_l + r_i}{2}\right)^{2(n-j)}$$

(3.19)

According to the different emphasis on social cognition and individual experience, r_g and r_i can be assigned different values. As an example of typical value settings: $r_g = r_i = r$, reflects equal emphasis on both types of information. In this case, Eqs. (3.18) and (3.19) can be simplified to:

$$E[Y_{(n)}] = (1 - r)^{n-j}$$

(3.20)

$$D[Y_{(n)}] = \left(1 - 2r + \frac{7r^2}{6}\right)^{n-j} - (1 - r)^{2(n-j)}$$

(3.21)

To ensure that the position update formula results in a bounded position vector, the mean and variance of $Y_{(n)}$, as described by Eqs. (3.20) and (3.21), must be bounded. Thus, the mean should not exceed 1, i.e., $0 \leq r \leq 2$, and the variance must not exceed 1, i.e., $|1 - 2r + 7r^2/6| \leq 1$, thus $0 \leq r \leq 1.7143$. Under these conditions, for any arbitrarily small non-negative ε, there exists an N such that:

$$\lim_{n \geq N} E[Y_{(n)}] = \lim_{n \geq N} (1 - r)^{n-j} < \varepsilon$$

(3.22)

$$\lim_{n \geq N} D[Y_{(n)}] = \lim_{n \geq N} \left[\left(1 - 2r + \frac{7r^2}{6}\right)^{n-j} - (1 - r)^{2(n-j)}\right] < \varepsilon$$

(3.23)

3.4.3 Analysis of the Weighted Parameter for the Initial Velocity Vector

Similar to the derivation of the factorial decay factor $Y_{(n)}$, the mathematical expectation of the weighted parameter $\delta^{(n)}$ for the initial velocity vector can be inferred from Eqs. (3.20) and (3.21).

$$E\left[w^j \cdot \prod_{m=j+1}^{n} \gamma(m)\right] = w^j \cdot (1-r)^{n-j-1}$$

$$E[\delta^{(n)}] = w^n + \sum_{j=1}^{n-1} E\left[\left(w^j \cdot \prod_{m=j+1}^{n} \gamma(m)\right)\right] = w^n + \sum_{j=1}^{n-1} w^j \cdot (1-r)^{n-j}$$

$$= \sum_{j=1}^{n} w^j \cdot (1-r)^{n-j} = w \cdot \frac{w^n - (1-r)^n}{w+r-1}$$

For typical parameter settings of the PSO algorithm, the velocity inertia factor w is generally a constant less than 1. Thus, we have:

$$\forall \varepsilon > 0, \exists N, \lim_{n \geq N} E[\delta^{(n)}] = \lim_{n \geq N} w \cdot \frac{w^n - (1-r)^n}{w+r-1} < \varepsilon \qquad (3.24)$$

This result indicates that the weighted parameter of the initial velocity vector $v_{id}^{(k)}$ will gradually converge to a region with a mean value of 0 as the number of iterations increases.

3.4.4 Discussion on the Similarity Between Standard PSO and Bare Bones PS

Based on the analysis from Eqs. (3.22) ~(3.24), after a particle has performed a certain number of search cycles based on $p_{ld}^{(k)}$ and $p_{id}^{(k)}$, we have:

$$\lim_{n \geq N}(Y_{(n)} \cdot p_{ld}^{(k)}) < \varepsilon \cdot p_{ld}^{(k)}$$

$$\lim_{n \geq N}(Y_{(n)} \cdot x_{id}^{(k)}) < \varepsilon \cdot x_{id}^{(k)}$$

$$\lim_{n \geq N}(\delta^{(n)} \cdot v_{id}^{(k)}) < \varepsilon \cdot v_{id}^{(k)}$$

Since $p_{ld}^{(k)}, x_{id}^{(k)}, v_{id}^{(k)}$ are bounded, the PSO update formula described by Eq. (3.17) can be simplified to:

$$x_{id}^{(k+n)} = p_{ld}^{(k)} + \phi^{(n)} \cdot \left(p_{id}^{(k)} - p_{ld}^{(k)}\right) \qquad (3.25)$$

According to the central limit theorem, under general conditions, as the number of independent random variables increases, their summation tends toward a normal distribution. When n is large enough, the random weight $\phi^{(n)}$ of the distance between $p_{ld}^{(k)}$ and $p_{id}^{(k)}$ tends toward a normal distribution.

BBPS [1] is a random sampling search process based on a normal distribution with a mean of $\frac{p_{ld}^{(k)} + p_{id}^{(k)}}{2}$ and a variance of $\left| p_{ld}^{(k)} - p_{id}^{(k)} \right|$, as defined by the following equation:

$$x_{id}^{(k+1)} = N\left(\frac{p_{ld}^{(k)} + p_{id}^{(k)}}{2}, \left| p_{ld}^{(k)} - p_{id}^{(k)} \right| \right) \tag{3.26}$$

BBPS can also be described by the following equation:

$$x_{id}^{(k+1)} = p_{ld}^{(k)} + N\left(\frac{p_{id}^{(k)} - p_{ld}^{(k)}}{2}, \left| p_{ld}^{(k)} - p_{id}^{(k)} \right| \right) \tag{3.27}$$

Kennedy [2] once pointed out, "BBPS doesn't really seem like a particle swarm. After all, we often describe particles as flying, whereas BBPS feels more like birds chirping rather than flying." Although the mathematical formulation of BBPS differs from classical PSO, a comparison between Eqs. (3.25) and (3.27), reveals that, standard PSO can be simplified to a normal distribution-based random sampling strategy under curtain conditions. This strategy is centered on the *gbest* position, with the distance between the *gbest* and the *pbest* particle as the radius. Thus, this mechanism is essentially the same as that described by Eq. (3.27) for BBPS. The difference is that BBPS simplifies the parameters, strictly defining the sampling principle of particles following a normal distribution and precisely determining the mean and variance of each particle in each cycle. This simplified form reduces parameter configuration the algorithm's execution and saves storage space in calculations.

3.5 Forgetting Characteristics of Parameters in a Probabilistic Sense

Lemma 3.3 provides a general expression for the movement patterns of particles. According to Eq. (3.13), this section analyzes the forgetting characteristics of standard PSO in a probabilistic sense and provides the value of the acceleration factor without $Vmax$ constraint, which is the Lipschitz condition of a particle dynamic system, to ensure the stability of particle motion. Considering Eq. (3.13), the maximum search space with single information for a standard PSO particle is as follows:

$$x_{id}^{(k+n)} = \left\{ \varphi_2^{(k+n)} + \sum_{j=1}^{n-1} \left(\varphi_2^{(k+j)} \cdot \prod_{m=j+1}^{n} \gamma(m) \right) \right\} \cdot p_{ld}^{(k)} + \left\{ \varphi_1^{(k+n)} + \sum_{j=1}^{n-1} \left(\varphi_1^{(k+j)} \cdot \right. \right.$$

$$\left. \left. \prod_{m=j+1}^{n} \gamma(m) \right) \right\} \cdot p_{id}^{(k)} + \left\{ \prod_{m=1}^{n} \gamma(m) \right\} \cdot x_{id}^{(k)} + \left\{ \omega^n + \sum_{j=1}^{n-1} \left(\omega^j \cdot \prod_{m=j+1}^{n} \gamma(m) \right) \right\} v_{id}^{(k)}$$

And it is expanded as shown in Fig. 3.4, each row corresponds to the summation series of four variables in the PSO algorithm: $p_{ld}^{(k)}, p_{id}^{(k)}, x_{id}^{(k)}$, and $v_{id}^{(k)}$. Notably, when analyzing by columns, each column corresponds to a particle position update equation for one cycle, and $\left\{ \prod_{m=j}^{n} \gamma(m) \right\}$ represents the weighted vector corresponding to that cycle (column). For example, in the first column, the weighted expression for $x_{id}^{(k+1)}$ is $\left\{ \prod_{m=j}^{n} \gamma(m) \right\}$:

$$x_{id}^{(k+1)} \Rightarrow \left\{ \varphi_{ld}^{(k+1)} \cdot p_{ld}^{(k)} + \varphi_{id}^{(k+1)} \cdot p_{id}^{(k)} + \gamma(1) \cdot x_{id}^{(k)} + \omega \cdot v_{id}^{(k)} \right\} \cdot \left\{ \prod_{m=2}^{n} \gamma(m) \right\}$$

$$= x_{id}^{(k+1)} \cdot \left\{ \prod_{m=2}^{n} \gamma(m) \right\}$$

Similarly, the j-th column's weighted expression for $x_{id}^{(k+j)}$ has a weight of $\left\{ \prod_{m=j+1}^{n} \gamma(m) \right\}$. Therefore, the position vector update equation at time $(k + n)$ can be described as follows:

$$x_{id}^{(k+n)} = x_{id}^{(k+1)} \cdot \left\{ \prod_{m=2}^{n} \gamma(m) \right\} + x_{id}^{(k+2)} \cdot \left\{ \prod_{m=3}^{n} \gamma(m) \right\} + \cdots + x_{id}^{(k+n-1)} \cdot \left\{ \gamma(n) \right\}$$

$$+ \left\{ \varphi_{ld}^{(k+n)} \cdot p_{ld}^{(k)} + \varphi_{id}^{(k+n)} \cdot p_{id}^{(k)} + \omega^n \cdot v_{id}^{(k)} \right\}$$

where the velocity inertia factor $|w|<1$, thus $\forall \varepsilon > 0, \exists N, \rightarrow \lim_{n>N} w^n < \varepsilon$, and the above equation can be expressed as Eq. (3.28).

$$x_{id}^{(k+n)} = \sum_{j=1}^{n-1} x_{id}^{(\pi)} \cdot \left\{ \prod_{m=j+1}^{n} \gamma(m) \right\} + \left\{ \varphi_{ld}^{(k+n)} \cdot p_{ld}^{(k)} + \varphi_{id}^{(k+n)} \cdot p_{id}^{(k)} \right\} \tag{3.28}$$

Equation (3.28) indicates that the current position vector is based on the random vectors of $p_{ld}^{(k)}$ and $p_{id}^{(k)}$, and the weighted sum of the position vectors from the previous $(k + n-1)$ cycles. The acceleration factor parameter values from Eqs. (3.20) and

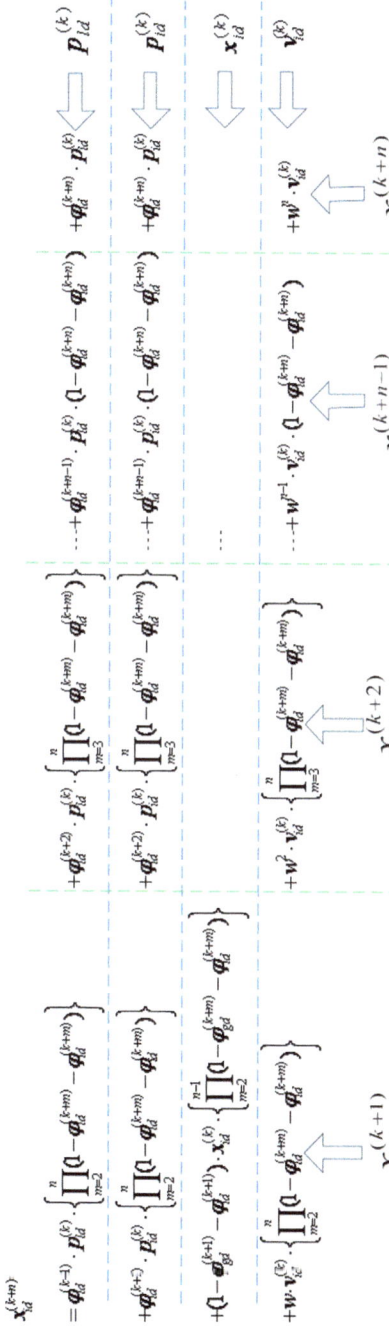

Fig. 3.4 Schematic diagram of the general description formula decomposition for $x_{id}^{(k)}$

(3.21) show that the weighted factor described by $\prod_{m=j}^{n} \gamma(m)$ exhibits a probabilistic forgetting characteristic. When the random values of the acceleration factor meet this condition, even without the *Vmax* constraint, it is ensured that the particles will search in a limited space, and with the increase of the search iteration, the weighted factor $Y = \prod_{m=j}^{n} \gamma(m)$ will also gradually decay (in a probabilistic sense) until the historical information (i.e., the previous position vectors) is completely forgotten. From the perspective of parameter selection, these results are consistent with Clerc's suggestions in the literature [3], where the value of the acceleration factor is 1.494 without the *Vmax* constraint, thereby expanding the acceptable range of of acceleration factor values.

Simulation experiments with different values of r are carried out. For each r value, 100,000 independent experiments are performed to statistically compare the number of search cycles n required for the weighted factor $Y = \prod_{m=j}^{n} \gamma(m)$ to decay below a threshold $\varepsilon(\varepsilon = 0.01)$, as shown in Fig. 3.5a and b. As seen in these figures, although the time for the weighted factor to decay below ε gradually increases with r, the peak distribution changes slowly and remains concentrated within 10 cycles. However, as r increases, the variance of the statistical results increases sharply. In Fig. 3.5d, while most decay cycles are concentrated within 75 cycles, but in some experiments, even after 350 cycles, the weighted factor remains above ε. This indicates that in Eq. (3.28), due to the high weighted coefficients corresponding to the past $(n-1)$ cycles, the weighted cumulative sum of the historical position vectors $x_{id}^{(k+j)}$ can far exceed the magnitudes of $p_{ld}^{(k+n)}$ and $p_{id}^{(k+n)}$, leading to individual behavior $x_{id}^{(k+n)}$ deviates from the guiding information (i.e., divergence).

3.6 Summary

Based on the mathematical model of the standard Particle Swarm Optimization (PSO) algorithm discussed in Chap. 2, this chapter simplifies the PSO algorithm into three distinct models and analyzes their algorithmic characteristics. The analysis shows that, without the introduction of new information, the *gbest* particle in the *Gbest* model will quickly reach a stagnation point and lose its ability to search further, which considerably restricts the search range of other particles. Therefore, it is necessary to modify the movement law of the *Gbest* particle to improve the algorithm. For example, Bergh's GCPSO introduces some randomness into the global best particle's update formula to change its movement mechanism.

Moreover, a general description for all particles is proposed, and the maximum search space with single information for standard PSO is summarized. Additionally, the chapter compares standard PSO with BBPS, analyzing their similarities. Finally, it presents the probabilistic forgetting characteristics of the parameters.

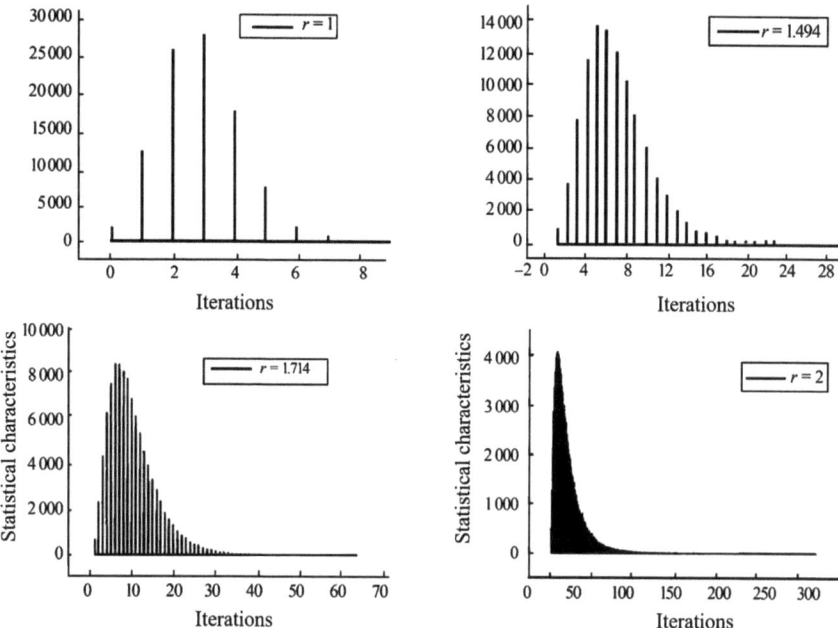

Fig. 3.5 Statistical characteristics Graph of $\prod\limits_{m=j}^{n} \gamma(m) < \varepsilon$

References

1. Qi J, Wang DW (2004) Particle swarm optimization algorithm for a model of optimally scheduling web advertising resources. Control Decis 8:881–884
2. Kennedy J (2003) Bare bones particle swarms. In: Proceedings of the 2003 IEEE swarm intelligence symposium. IEEE, Piscataway, NJ, USA
3. Krink T, Vesterstrom JS, Riget J (2002) Particle swarm optimisation with spatial particle extension. In: Proceedings of the 2002 world congress on computational intelligence—WCCI'02. IEEE, Piscataway, NJ, USA

Chapter 4
Sampling Distribution and Particle Trajectories in Standard PSO

Over the past decade, various theoretical analyses of particle swarm dynamics have emerged in the literature. Ozcan and Mohan [1] examined the behavior of a particle in an isolated, one-dimensional space, without randomness and in a state of stagnation. Similarly, p_{id} and p_{ld} were assumed to be identical, which can occur for the best particle in a neighborhood. In Ozcan and Mohan [2], Al-Kazemi and Mohan [3], Ozcan and Mohan extended this study to multidimensional spaces, analyzing the coverage of multiple particles under similar conditions. The same assumptions were applied in the model by Clerc and Kennedy [4]: a single particle in a one-dimensional space, deterministic behavior, and stagnation. The dynamics of the particle state (position and velocity) were determined by calculating the eigenvalues and eigenvectors of the transition matrix. As a result, the model predicted that the particle would converge to an equilibrium state (if the eigenvalue was less than 1).

A similar approach was used by van den Bergh [5], focusing on a single particle without randomness and in a state of stagnation. In his earlier research, he proposed an exact solution to simulate a particle's trajectory. He suggested that the particle is attracted to a fixed point and demonstrated that this analysis still holds even when randomness is introduced. It was further noted in Van den Bergh [5, 6] that the fixed point to which the particle converges might not necessarily correspond to either the *gbest* or *lbest*.

Yasuda et al. investigated a simplified PSO model [7], assuming a one-dimensional particle without randomness, and in a stagnant state. By analyzing the eigenvalues of the dynamic system, they determined the parameter settings necessary to ensure system stability and predict possible particle behavior. The paper provides a detailed analysis of "ring" behavior within the model, which applies an inertia weight.

Blackwell [8] examined how a particle swarm evolves over iterations under spatial constraints. The simplified swarm model extends the work of *Clerc and Kennedy*, incorporating multiple particles and multidimensional space. The model allows for interactions between particles and enables them to update their pbest. Although a constriction coefficient is used, the model does not include randomness. *Blackwell* also indicates that the spatial expansion decreases exponentially over time.

Brandstatter and Baumgartner [9] compared *Clerc and Kennedy's* model to a damped mass-spring oscillator, reformulating it from the perspective of damping factors and natural oscillation frequency. Like the original model, this one assumes a single particle in one-dimensional space, without randomness and in a stagnant state.

Following the same assumptions as in Clerc and Kennedy [4] and using a similar approach, Trelea [10] provided a thorough analysis of a particle model with four parameters, identifying regions within the parameter space and qualitatively analyzing different behaviors of the model (e.g., stability, harmonic oscillation, or zigzag motion).

The dynamic system proposed by *Clerc* and *Kennedy* has been further extended by Campana et al. [13, 14], where they explored an extended PSO model. Under their assumptions, the current state lacks randomness, resulting in a discrete and linear fixed dynamic system. Jinyu et al. [11], Wen et al. 12], Campana et al. 13, 14] describe both unforced and forced responses. However, because the forced response depends on the characteristics of the fitness function, only the natural response can be analyzed in detail.

To gain a better understanding of PSO behavior during stagnation, Clerc [15] analyzed the velocity distribution of a particle in the standard PSO, which includes inertia weight and stochastic constraints in its update rules. He specifically demonstrated that a particle's new velocity is composed of three components: a forward force, an opposing force, and noise. *Clerc* investigated the distribution of these forces.

Kadirkamanathan et al. [16] employed *Lyapunov* stability analysis to study the stability of particles in the presence of stochasticity. They examined the behavior of a particle—the *gbest* particle—under inertia weight and stagnation conditions. They derived theoretical insights from control theory by modeling the particle as a nonlinear feedback system. For example, they found that PSO parameters can ensure convergence under certain conditions. Because *Lyapunov* theory is quite conservative, the conditions identified are stringent, effectively ensuring that PSO exhibits minimal oscillation.

Poli et al. [17] assumed that particles are in a stagnation phase and analyzed the distribution of particle positions to accurately determine the properties of the sampling distribution in PSO. They also explained how these characteristics evolve during iterations and examined the generation of sampling distributions as particles seek their *gbest* positions.

The purpose of this chapter is to discuss the theoretical analysis of particle trajectories and the sampling distributions and to summarize the results of these studies. Section 4.1 reviews Poli's analysis [17] regarding the sampling distribution characteristics of PSO and its convergence during stagnation. Section 4.2, drawing on the ideas of Bergh [5], derives particle trajectories through position analysis.

4.1 Sampling Distribution and Convergence During Stagnation in Standard PSO

In Poli et al. [17], a model (4.1) is introduced to analyze the sampling distribution and convergence properties of the standard PSO during stagnation. When the PSO algorithm is in a state of stagnation (i.e., no better fitness values are found), each particle's behavior becomes independent, with both the pariticle's *pbest* ($p_i^{(k)}$) and the *gbest* ($p_l^{(k)}$) remaining constant. Consequently, the analysis focuses independently on a single particle in each dimension. Under these conditions, the update equations for velocity and position in PSO can be reformulated as follows:

$$\begin{cases} v_{id}^{(k)} = x_{id}^{(k)} - x_{id}^{(k-1)} \\ x_{id}^{(k+1)} = (1+w)x_{id}^{(k)} - \left(\varphi_1^{(k+1)} + \varphi_2^{(k+1)}\right)x_{id}^{(k)} - wx_{id}^{(k-1)} + \varphi_1^{(k+1)}p_{id}^{(k)} + \varphi_2^{(k+1)}p_{ld}^{(k)} \end{cases}$$

(4.1)

As defined in Chap. 3, $x_{id}^{(k+1)}$ represents the position vector of the i-th particle in the d-th dimension at the $k+1$ cycle. All other variables adhere to the same subscript and superscript conventions defined in Chap. 3.

Similar to the starting point of Sect. 3.3, Poli and Broomhead [17] also establishes a new statistical dynamics model for PSO under the stagnation hypothesis and analyzes its sampling distribution characteristics. The objective of this study closely aligns with Lemma 3.3.

The research logic of Poli and Broomhead [17] is as follows:

(1) **From Expectation to Higher-Order Statistics:**

Starting from the position update equation, expectation operators are applied to derive recurrence relations for the mean $E\left[x_{id}^{(k)}\right]$, second-order moment $E\left[(x_{id}^{(k+1)})^2\right]$, and covariance $E\left[x_{id}^{(k)} \cdot x_{id}^{(k-1)}\right]$.

(2) **Unified Framework for Stability Analysis:**

The recurrence equations are transformed into a state-space model $z(k+1) = Mz(k) + b$, enabling unified convergence analysis of all statistics via matrix eigenvalues.

(3) **Separated Convergence Conditions:**

System stability depends solely on parameters, independent of the positions of *pbest* $p_{id}^{(k)}$ and *gbest* $p_{ld}^{(k)}$.

Based on the standard PSO algorithm (without loss of generality), Poli and Broomhead [17] adopts the following assumptions:

(1) **Stagnation Hypothesis**: Particles fail to find better solutions (the same as Chap. 3), with both $p_{id}^{(k)}$ and $p_{ld}^{(k)}$ remaining constant. Each dimension is analyzed independently.

(2) **Independence of Stochastic Variables**: The random coefficients $\phi_1^{(k+1)}$ and $\phi_2^{(k+1)}$ are independent of historical positions $x_{id}^{(k+1)}$ and follow uniform distributions $(E\left[\phi_i^{(k+1)}\right] = \frac{c}{2}, E\left[\left(\phi_i^{(k+1)}\right)^2\right] = \frac{c^2}{3})$.

(3) **Linear System Assumption**: Dynamic equations can be expressed as linear recursions or in a matrix form, facilitating stability analysis via eigenvalues.

4.1.1 Dynamic Equation for $E\left[x_{id}^{(k)}\right]$

This section derives the recurrence equation for the mean of particle positions $E\left[x_{id}^{(k)}\right]$, establishes a dynamic model for the first-order statistical moments, and analyzes their fixed points and stability conditions. Using the position update equation Eq. (4.1), the mean of both sides is:

$$
E\left[x_{id}^{(k+1)}\right] = (1+w) \cdot E\left[x_{id}^{(k)}\right] - E\left[\left(\varphi_1^{(k+1)} + \varphi_2^{(k+1)}\right)x_{id}^{(k)}\right] - w \cdot E\left[x_{id}^{(k-1)}\right]
$$
$$
+ E\left[\varphi_1^{(k+1)}p_{id}^{(k)}\right] + E\left[\varphi_2^{(k+1)}p_{ld}^{(k)}\right]
\tag{4.2}
$$

Considering the aforementioned assumption, we have:

$$
E\left[x_{id}^{(k+1)}\right]
\begin{aligned}
&= (1+w) \cdot E\left[x_{id}^{(k)}\right] - E\left[\varphi_1^{(k+1)}\right] \cdot E\left[x_{id}^{(k)}\right] + E\left[\varphi_2^{(k+1)}\right] \cdot E\left[x_{id}^{(k)}\right] \\
&\quad -w \cdot E\left[x_{id}^{(k-1)}\right] + p_{id}^{(k)} \cdot E\left[\varphi_1^{(k+1)}\right] + p_{id}^{(k)} \cdot E\left[\varphi_2^{(k+1)}\right] \\
&= (1+w-c) \cdot E\left[x_{id}^{(k)}\right] - w \cdot E\left[x_{id}^{(k-1)}\right] + c\frac{p_{id}^{(k)}+p_{ld}^{(k)}}{2}
\end{aligned}
\tag{4.3}
$$

The characteristic equation of Eq. (4.3) can be expressed as:

$$
\lambda^2 - (1+w-c) \cdot \lambda + \omega = 0
\tag{4.4}
$$

Thus, we can obtain that the stability conditions for the mean subsystem in Eq. (4.3) is given by $|w| < 1$ and $0 < c < 2 + 2w$. Notably, in contrast to [17], where the value of w is restricted $0 < w < 1$, our conclusion extends the permissible range of w to $(-1, 0]$. This analysis yields the same conclusion as that reached in Chap. 5. Moreover, when $w \leq 0$, the upper bound for c falls below 2, indicating that negative inertia (oscillatory reverse memory) is indeed feasible, although the allowable interval for c contracts significantly. Figure 4.1 illustrates the relationship between the eigenvalues and the parameters w and c.

Fig. 4.1 Stability region of
$E\left[x_{id}^{(k)}\right]$

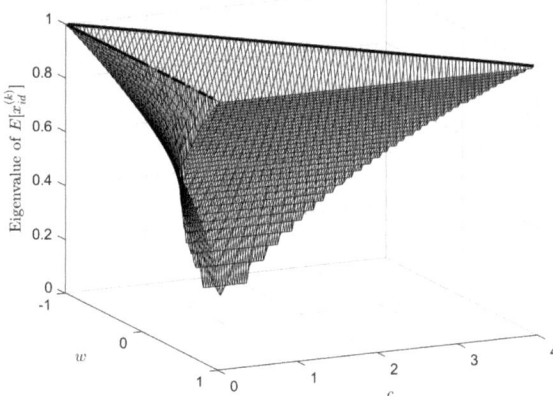

4.1.2 Dynamic Equations for $E\left[\left(x_{id}^{(k+1)}\right)^2\right]$, $E\left[x_{id}^{(k)}\cdot x_{id}^{(k-1)}\right]$ and $D\left[x_{id}^{(k)}\right]$

According to the aforementioned independence assumption, the core idea in the subsequent derivation is to decompose higher-order statistical moments into linear combinations of lower-order terms. By expanding the position update equation, dynamic equations for the second-order moments and covariances are established, thereby enabling an investigation into the dispersion characteristics of the sampling distribution (e.g., whether the standard deviation converges). Based on Eqs. (4.1), (4.2) and (4.3), the second-order moments $E\left[(x_{id}^{(k+1)})^2\right]$ are calculated as follows:

$$
\begin{aligned}
E\left[(x_{id}^{(k+1)})^2\right] =\ & E\left[\left((1+w-\varphi_1^{(k+1)}-\varphi_2^{(k+1)})\cdot x_{id}^{(k)}-w\cdot x_{id}^{(k-1)}+\varphi_1^{(k+1)}p_{id}^{(k)}+\varphi_2^{(k+1)}p_{ld}^{(k)}\right)^2\right]\\
=\ & E\left[((1+w-\varphi_1^{(k+1)}-\varphi_2^{(k+1)})\cdot x_{id}^{(k)})^2\right]\\
& +E\left[-2w(1+w-\varphi_1^{(k+1)}-\varphi_2^{(k+1)})\cdot x_{id}^{(k)}x_{id}^{(k-1)}\right]\\
& +E\left[2(1+w-\varphi_1^{(k+1)}-\varphi_2^{(k+1)})\cdot(\varphi_1^{(k+1)}p_{id}^{(k)}+\varphi_2^{(k+1)}p_{ld}^{(k)})\cdot x_{id}^{(k)}\right]\\
& +E\left[(w\cdot x_{id}^{(k-1)})^2\right]+E\left[-2w(\varphi_1^{(k+1)}p_{id}^{(k)}+\varphi_2^{(k+1)}p_{ld}^{(k)})\cdot x_{id}^{(k-1)}\right]\\
& +E\left[(\varphi_1^{(k+1)}p_{id}^{(k)}+\varphi_2^{(k+1)}p_{ld}^{(k)})^2\right]\\
E\left[(x_{id}^{(k+1)})^2\right] =\ & E\left[\left(x_{id}^{(k)}\right)^2 2c(1+w)+\tfrac{7}{6}c^2\right]\cdot\left((1+w)^2-\right)+E\left[x_{id}^{(k)}x_{id}^{(k-1)}\right]\cdot(-2w(1+w-c))\\
& +E\left[x_{id}^{(k)}\right]\cdot\left(c\left(1+w-\tfrac{7}{6}c\right)p_{id}^{(k)}+c\left(1+w-\tfrac{7}{6}c\right)p_{ld}^{(k)}\right)+E\left[(x_{id}^{(k-1)})^2\right]\cdot w^2\\
& +E\left[x_{id}^{(k-1)}\right]\cdot\left(-wc\left(p_{id}^{(k)}+p_{ld}^{(k)}\right)\right)+\tfrac{1}{3}c^2\left(\left(p_{id}^{(k)}\right)^2+\tfrac{3}{2}p_{id}^{(k)}p_{ld}^{(k)}+\left(p_{ld}^{(k)}\right)^2\right)
\end{aligned}
$$

$$(4.5)$$

where $\mu = E\left[\varphi_i^{(k+1)}\right]=\frac{c}{2}$, $v = E\left[\varphi_i^{(k+1)^2}\right]=\frac{c^2}{3}$.

Based on Eq. (4.1), the product $x_{id}^{(k)} \cdot x_{id}^{(k-1)}$ is calculated as:

$$x_{id}^{(k)} \cdot x_{id}^{(k-1)} = (1+w)\left(x_{id}^{(k-1)}\right)^2 - \left(\varphi_1^{(k)} + \varphi_2^{(k)}\right)\left(x_{id}^{(k-1)}\right)^2 - wx_{id}^{(k)}x_{id}^{(k-1)}$$
$$+ \varphi_1^{(k)}p_{id}^{(k)}x_{id}^{(k-1)} + \varphi_2^{(k)}p_{ld}^{(k)}x_{id}^{(k-1)} \tag{4.6}$$

Then we have:

$$E\left[x_{id}^{(k)} \cdot x_{id}^{(k-1)}\right] = (1+w-c) \cdot E\left[(x_{id}^{(k-1)})^2\right] - w \cdot E\left[x_{id}^{(k)}x_{id}^{(k-1)}\right]$$
$$+ c\frac{p_{id}^{(k)} + p_{ld}^{(k)}}{2}E\left[x_{id}^{(k-1)}\right] \tag{4.7}$$

With the previously calculated $E\left[x_{id}^{(k+1)}\right]$ and $E\left[(x_{id}^{(k+1)})^2\right]$, the standard deviation of the sampling distribution when PSO is in a stagnation state can be derived as:

$$D\left[x_{id}^{(k)}\right] = \sqrt{E\left[\left(x_{id}^{(k)}\right)^2\right] - \left(E\left[x_{id}^{(k)}\right]\right)^2} \tag{4.9}$$

In summary, the recursions for $E\left[x_{id}^{(k+1)}\right]$, $E\left[(x_{id}^{(k+1)})^2\right]$, and $E\left[x_{id}^{(k)} \cdot x_{id}^{(k-1)}\right]$ are as follows:

$$\begin{cases} E\left[x_{id}^{(k+1)}\right] & = (1+w-c) \cdot E\left[x_{id}^{(k)}\right] - w \cdot E\left[x_{id}^{(k-1)}\right] + c\frac{p_{id}^{(k)} + p_{ld}^{(k)}}{2} \\ E\left[(x_{id}^{(k+1)})^2\right] & = E\left[\left(x_{id}^{(k)}\right)^2\right] \cdot ((1+w)^2 - 2c(1+w) + \frac{7}{6}c^2) \\ & + E\left[x_{id}^{(k)}x_{id}^{(k-1)}\right] \cdot (-2w(1+w-c)) \\ & + E\left[x_{id}^{(k)}\right] \cdot ((c(1+w) - \frac{7}{6}c^2)p_{id}^{(k)} + (c(1+w) - \frac{7}{6}c^2)p_{ld}^{(k)}) \\ & + E\left[(x_{id}^{(k-1)})^2\right] \cdot w^2 \\ & + E\left[x_{id}^{(k-1)}\right] \cdot (-wc(p_{id}^{(k)} + p_{ld}^{(k)})) \\ & + \frac{1}{3}c^2((p_{id}^{(k)})^2 + \frac{3}{2}p_{id}^{(k)}p_{ld}^{(k)} + (p_{ld}^{(k)})^2) \\ E\left[x_{id}^{(k)} \cdot x_{id}^{(k-1)}\right] & = (1+w-c) \cdot E\left[(x_{id}^{(k-1)})^2\right] - w \cdot E\left[x_{id}^{(k)}x_{id}^{(k-1)}\right] \\ & + c\frac{p_{id}^{(k)} + p_{ld}^{(k)}}{2}E\left[x_{id}^{(k-1)}\right] \end{cases}$$
$$\tag{4.10}$$

4.1.3 Stability Analysis of Particles with Randomness

The system described by Eq. (4.10) can be rewritten in the form of a state equation:

$$z(k + 1) = Mz(k) + b \qquad (4.11)$$

where $z(k) = \left[E(x^{(k)})\ E(x^{(k-1)})\ E((x^{(k)})^2)\ E((x^{(k-1)})^2)\ E(x^{(k)}x^{(k-1)}) \right]^T$. For simplicity, let $p_{id}^{(k)} = p_{ld}^{(k)} = p$.

$$M = \begin{bmatrix} 1 + w - c & -w & 0 & 0 & 0 \\ 1 & 0 & 0 & 0 & 0 \\ 2cp(1 + w - \frac{7}{6}c) & -2wcp & (1+w)^2 - 2c(1+w) + \frac{7}{6}c^2 & w^2 & -2w(1+w-c) \\ 0 & 0 & 1 & 0 & 0 \\ cp & 0 & 1 + w - c & 0 & -w \end{bmatrix}$$

$$b = \left[cp\ 0\ \tfrac{7}{6}c^2p^2\ 0\ 0 \right]^T$$

The characteristic polynomial of the matrix M can be simplified by exploiting its block structure. Note that the evolution of $E\left[x_{id}^{(k)}\right]$ is independent of the second-order moments (indeed, in the expression for M, the first two columns act solely on the first-moment subspace). Hence, M can be regarded as a lower triangular block matrix coupling a mean subsystem (the upper 2×2 block) with a second-order-moment subsystem (the lower 3×3 block).

The mean subsystem is described by the difference equation in Eq. (4.3), whose homogeneous characteristic equation is given by Eq. (4.4). The solutions are:

$$\lambda_{1,2} = \frac{1 + \omega - c \pm \sqrt{(\omega - c)^2 - 2c - 2\omega + 1}}{2}$$

which are the two simple eigenvalues mentioned in [17]. The remaining three eigenvalues belong to the second-order-moment subsystem; their analytical expressions are exceedingly complicated and, notably, are independent of $p_{id}^{(k)}$ and $p_{ld}^{(k)}$. For example, by substituting the parameter values $w = 0.7298$ and $c = 1.49618$ suggested in [4], one obtains the two simple eigenvalues approximately as: $\lambda_{1,2} \approx 0.1168 \pm 0.8463i$. Since the moduli of these eigenvalues are less than 1, they correspond to a damped oscillatory convergence mode of the mean. The remaining three eigenvalues of M are 0.8927 and $-0.3310 \pm 0.5708i$. Among these, one eigenvalue is real while the other two occur as a complex conjugate pair, corresponding to the convergence modes of the variance and covariance. With all eigenvalue moduli residing within the unit circle, the overall system is stable.

Figure 4.1 illustrates one of the necessary conditions that can be readily obtained from the characteristic equation of the mean subsystem; however, this condition alone does not suffice to guarantee overall stability. To further ensure convergence of the second-order-moment subsystem, we delineated the stability region of the spectral radius of M via numerical simulation. As shown in Fig. 4.2, the overall stability region is strictly contained within the mean stability region, which corroborates that the requirements imposed by the Lyapunov-Schur criterion are more stringent than those needed for mere convergence of the mean.

The analysis above reveals a significant phenomenon: there exists a class of parameter combinations for which the first moment (mean) of the particle positions converges, while the second moment (variance) persistently diverges. In particular, parameters that satisfy mean stability yet fail to ensure overall stability—for example, for $w = 0.8$, the range $1.557 < c < 2(1 + 0.8) = 3.6$—falls into this category. In this case, the eigenvalues of the mean subsystem (i.e., $\rho(M_{1\times1}) < 1$ guarantee that $E\left[x_{id}^{(k)}\right] \to p$; however, the second-order-moment subsystem has eigenvalues that are greater than or equal to 1, leading to divergence of the variance and standard deviation (manifesting as either linear drift or exponential growth). From the viewpoint of the eigenstructure, this implies that the spectrum of M contains an eigenvalue equal to or very close to 1 (corresponding to the variance dimension), i.e., $\rho(M) > 1$, even though the remaining eigenvalues lie strictly within the unit circle. Moreover, Poli and Broomhead [17] pointed out that when parameters are chosen between the two stability boundary curves, $E\left[x_{id}^{(k)}\right]$ converges to p, while $D\left[x_{id}^{(k)}\right]$ "slowly" diverges to infinity. Under such circumstances, although the particles oscillate around the mean position, the amplitude of these oscillations gradually increases—which, for the PSO algorithm, might actually facilitate the particles escaping from local minima.

Fig. 4.2 Stability region comparison of $E\left[x_{id}^{(k)}\right]$ and M

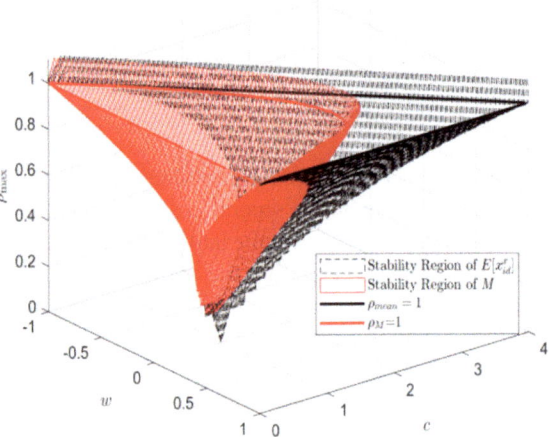

4.2 Position Analysis of Particle Trajectories

Liu [18] conducted an analysis of the particle trajectories in PSO. The trajectory analysis is the same for different particles in different dimensions. The update equations of the standard PSO are rewrite as follows:

$$\begin{cases} v_{id}^{(k+1)} = w \cdot v_{id}^{(k)} + c_1 \cdot r_1 \cdot (p_{id}^{(k)} - x_{id}^{(k)}) + c_2 \cdot r_2 \cdot (p_{ld}^{(k)} - x_{id}^{(k)}) \\ x_{id}^{(k+1)} = x_{id}^{(k)} + v_{id}^{(k+1)} \end{cases} \quad (4.12)$$

where $\varphi_1 = c_1 \cdot r_1$, $\varphi_2 = c_2 \cdot r_2$, and $\varphi = \varphi_1 + \varphi_2$, while $p_{id}^{(k)}$, $p_{ld}^{(k)}$, φ_1, and φ_2 remain constant, Eq. (4.12) can be formulated as:

$$x_{id}^{(k+1)} = (1 + \omega - \varphi) \cdot x_{id}^{(k)} - w \cdot x_{id}^{(k-1)} + \varphi_1 p_{id}^{(k)} + \varphi_2 p_{ld}^{(k)} \quad (4.13)$$

The corresponding homogeneous matrix form of Eq. (4.13) is:

$$\begin{bmatrix} x_{id}^{(k+1)} \\ x_{id}^{(k)} \\ 1 \end{bmatrix} = A \cdot \begin{bmatrix} x_{id}^{(k)} \\ x_{id}^{(k-1)} \\ 1 \end{bmatrix}, A = \begin{bmatrix} 1 + w - \varphi & -w & \varphi_1 \cdot p_{id}^{(k)} + \varphi_2 \cdot p_{ld}^{(k)} \\ 1 & 0 & 0 \\ 0 & 0 & 1 \end{bmatrix} \quad (4.14)$$

The characteristic polynomial of the coefficient matrix A in Eq. (4.14) is:

$$(\lambda - 1)(\lambda^2 - (1 + w - \varphi)\lambda + w) = 0 \quad (4.15)$$

Equation (4.15) has three eigenvalues. In addition $\lambda = 1$, the remaining two eigenvalues are given by:

$$\begin{aligned} \alpha &= \frac{1 + \omega - \varphi + \sqrt{(1 + w - \varphi)^2 - 4w}}{2} \\ \beta &= \frac{1 + \omega - \varphi - \sqrt{(1 + w - \varphi)^2 - 4w}}{2} \end{aligned} \quad (4.16)$$

Therefore, Eq. (4.16) can be expressed as:

$$x_{id}^{(k)} = b_1 + b_2 \alpha^k + b_3 \beta^k \quad (4.17)$$

In Eq. (4.17), b_1, b_2, and b_3 are constants determined by the system's initial conditions. In order to determine these three constants, only three equations are needed. Based on Eq. (4.13), if $x_{id}^{(k)}$ and $x_{id}^{(k+1)}$ are given, $x_{id}^{(k+2)}$ can be calculated. With these three initial conditions, the following equations can be derived from Eq. (4.17):

$$
\begin{bmatrix} x_{id}^{(k)} \\ x_{id}^{(k+1)} \\ x_{id}^{(k+2)} \end{bmatrix} = \begin{bmatrix} 1 & 1 & 1 \\ 1 & \alpha & \beta \\ 1 & \alpha^2 & \beta^2 \end{bmatrix} \cdot \begin{bmatrix} b_1 \\ b_2 \\ b_3 \end{bmatrix} \tag{4.18}
$$

Solving the system of Eq. (4.18) and using Eq. (4.16) yields:

$$
\begin{aligned}
b_1 &= \frac{\varphi_1 \cdot p_{id}^{(k+1)} + \varphi_2 \cdot p_{ld}^{(k+1)}}{\varphi_1 + \varphi_2} \\
b_2 &= \frac{\alpha(x_{id}^{(k+1)} - x_{id}^{(k)}) + x_{id}^{(k+1)} - x_{id}^{(k+2)}}{\gamma(\beta - 1)} \\
b_3 &= \frac{\beta(x_{id}^{(k+1)} - x_{id}^{(k)}) + x_{id}^{(k+1)} - x_{id}^{(k+2)}}{\gamma(\alpha - 1)}
\end{aligned} \tag{4.19}
$$

We can see from Eq. (4.17) that for $x_{id}^{(k)}$ to converge, the absolute values of α and β must be less than 1 (as determined by Eq. (4.16)). Therefore, if $max(\|\alpha\|, \|\beta\|) < 1$, the particle's position trajectory $x_{id}^{(k)}$ converges according to the following equation:

$$
\lim_{k \to \infty} x(k) = (1 - a)p_{id}^{(k+1)} + ap_{ld}^{(k+1)} \qquad a = \frac{\varphi_1}{\varphi_1 + \varphi_2} \tag{4.20}
$$

For Eq. (4.20) to hold, the condition $max(\|\alpha\|, \|\beta\|) < 1$ must be satisfied. Given this inequality, we can derive constraints on w, φ_1, and φ_2 that ensure Eq. (4.20) remains valid—that is, to ensure convergence of the particle's trajectory. From Eq. (4.16), two cases can be considered:

(1) **Complex eigenvalues**: When $(1 + w - \varphi)^2 < 4w$, α and β are complex numbers, then:

$$
\|\|\alpha\| = \|\beta\| = \frac{1}{2}\sqrt{(1 + w - \varphi)^2 + 4\omega - (1 + w - \varphi)^2} = \sqrt{w}
$$

If $w < 1$ and $(1 + w - \varphi)^2 < 4w$, Eq. (4.20) holds. Therefore, under these conditions, the particle's trajectory converges.

(2) **Real eigenvalues**: When $(1 + w - \varphi)^2 \geq 4w$, α and β are real numbers. From Eq. (4.16), we deduce that if $1 + w - \varphi > 0$, then $\|\alpha\| > \|\beta\|$; and if $1 + w - \varphi \leq 0$, then $\|\alpha\| \leq \|\beta\|$. In this scenario, two sub-cases arise:

① When $1 + w - \varphi > 0$ and $\|\alpha\| > \|\beta\|$, then $\|\alpha\| < 1$, and thus the system converges. Based on Eq. (4.16), $\|\alpha\| < 1$ implies $0 < \frac{1 + w - \varphi + \sqrt{(1+w-\varphi)^2 - 4w}}{2} < 1$, which leads to $\varphi > 0$.

② When $1 + \omega - \varphi < 0$ and $\|\alpha\| \leq \|\beta\|$ (i.e., $\|\beta\| < 1$), the system converges. Based on Eq. (4.16), $\|\beta\| < 1$ implies $0 < |\frac{1 + w - \varphi - \sqrt{(1+w-\varphi)^2 - 4w}}{2}| < 1$, which leads to $w > \frac{1}{2}\varphi - 1$.

In summary, the conditions for the convergence of particle trajectory are given by either (4.21), (4.22), or (4.23).

$$\begin{cases} \omega < 1 \\ (1 + \omega - \varphi)^2 < 4\omega \end{cases} \tag{4.21}$$

or

$$\begin{cases} \varphi > 0 \\ 1 + \omega - \varphi > 0 \\ (1 + \omega - \varphi)^2 \geq 4\omega \end{cases} \tag{4.22}$$

or

$$\begin{cases} \omega > \dfrac{1}{2}\varphi - 1 \\ 1 + \omega - \varphi < 0 \\ (1 + \omega - \varphi)^2 \geq 4\omega \end{cases} \tag{4.23}$$

Therefore, when randomness is not considered, and *pbest* and the *gbest* are constants, the convergence condition for the PSO algorithm (illustrated in Fig. 4.3) is given by Eq. (4.24).

$$\begin{cases} 0 < \omega < 1 \\ 0 < \varphi < 4 \\ \omega > \dfrac{1}{2}\varphi - 1 \end{cases} \tag{4.24}$$

4.3 Summary

This chapter presents the theoretical analysis of particle trajectories and sampling distribution and summarizes the findings from these studies. Firstly, an analysis of the sampling distribution of the standard PSO and its convergence under stagnation is provided. The dynamic equations for $E\left[x_{id}^{(k)}\right]$, $E\left[(x_{id}^{(k+1)})^2\right]$, $E\left[x_{id}^{(k)} \cdot x_{id}^{(k-1)}\right]$, and $D\left[x_{id}^{(k)}\right]$ are calculated, describing the PSO system as a state equation. It is found that $E(x_k^2)$ converges to p^2 as k approaches infinity. Let $E(x_k)$ converge to p (without ensuring that $D\left[x_{id}^{(k)}\right]$ converges to 0), only when $p = p_l$ does $E(x_k^2)$ converge to p^2 as k approaches infinity. In other words, PSO will stop searching only when $p = p_l$. For particle convergence, it is required not only that $\lim_{k \to \infty} E(x_k) = p$ but

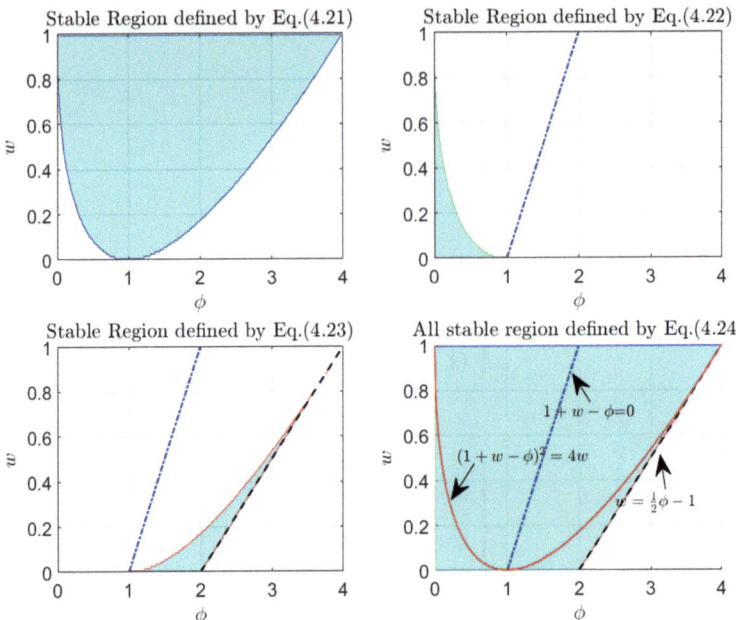

Fig. 4.3 The Stable Region of coefficient matrix A in Eq. (4.14)

also that $\lim_{k \to \infty} stdDev(x_k) = 0$. Next, the position trajectory of particles is analyzed and discussed, and trajectory convergence requires three conditions, as shown in Eqs. (4.21), (4.22) and (4.23).

This section synthesizes two complementary views of PSO convergence. Section 4.1 shows, via a state-space, moment-based analysis under the stagnation hypothesis, that a particle's mean position converges only when the algorithm's parameters keep all eigenvalues of the mean–variance system within the unit circle, conditions that depend solely on w, c_1, c_2. Section 4.2 arrives at the same requirement from a deterministic trajectory perspective: the characteristic eigenvalues α, β of the position-update matrix must satisfy $\|\alpha\|, \|\beta\| < 1$, yielding explicit inequalities (Eqs. (4.21)~(4.24) that define a "stable region" for w, φ. Together, these results establish that suitable parameter choices guarantee damped oscillations that converge to an equilibrium (full stagnation), whereas settings outside the stable region cause persistent oscillations or divergence, sometimes useful for escaping local minima but forfeiting convergence guarantees. Thus, PSO's behavior is fundamentally a trade-off governed by its coefficients: tighter stability bounds ensure convergence; looser bounds sustain exploration.

References

1. Ozcan E, Mohan CK (1998) Analysis of a simple particle swarm optimization system. In: Proceedings of the 1998 artificial networks in engineering conference, ANNIE, November 1, 1998–November 4, 1998. ASME, St.Louis, MO, USA
2. Ozcan E, Mohan CK (1999) Particle swarm optimization: surfing the waves. In: Proceedings of the 1999 congress on evolutionary computation-CEC99. IEEE, Piscataway, NJ, USA
3. Al-Kazemi B, Mohan CK (2002) Multi-phase generalization of the particle swarm optimization algorithm. In: Proceedings of 2002 world congress on computational intelligence—WCCI'02. IEEE, Piscataway, NJ, USA
4. Clerc M, Kennedy J (2002) The particle swarm—explosion, stability, and convergence in a multidimensional complex space [J]. IEEE Trans Evol Comput 6(1):58–73
5. van den Bergh F (2002) An analysis of particle swarm optimizers [D]. University of Pretoria, South Africa
6. van den Bergh F, Engelbrecht AP (2006) A study of particle swarm optimization particle trajectories [J]. Inf Sci 176(8):937–971
7. Iwasaki N, Yasuda K (2005) Adaptive particle swarm optimization via velocity feedback [J]. Trans Inst Electric Eng Japan Part C 125-C(6):987–988
8. Blackwell TM (2005) Particle swarms and population diversity. Soft Comput 9(11):793–802
9. Brandstatter B, Baumgartner U (2002) Particle swarm optimization—mass-spring system analogon. Institute of Electrical and Electronics Engineers Inc.
10. Trelea IC (2003) The particle swarm optimization algorithm: convergence analysis and parameter selection [J]. Inf Process Lett 85(6):317–325
11. Jinyu W, Shaorong W, Shijie C, et al (1998) Measurement based power system load modeling using a population diversity genetic algorithm. In: Proceedings of the 1998 International Conference on power system technology, POWERCON '98
12. Wen JY, Wu QH, Shimmin DW, et al (1999) Population diversity based genetic algorithm for fuzzy control of synchronous generators. In: Proceedings of the IEEE international symposium on computer aided control system design
13. Campana E, Fasano G, Pinto A (2006) Dynamic system analysis and initial particles position in particle swarm optimization. In: Proceedings of the IEEE swarm intelligence symposium, Indianapolis
14. Campana E, Fasano GDP, et al (2006) Particle swarm optimization: effcient globally convergent modifications. In: Proceedings of the the III European conference on conputational mechanics, solids, structures and coupled problems in engineering, Lisbon, Portugal
15. Clerc M (2006) Stagnation analysis in particle swarm optimization or what happens when nothing happens
16. Kadirkamanathan V, Selvarajah K, Fleming PJ (2006) Stability analysis of the particle dynamics in particle swarm optimizer [J]. IEEE Trans Evol Comput 10(3):245–255
17. Poli R, Broomhead D (2007) Exact analysis of the sampling distribution for the canonical particle swarm optimiser and its convergence during stagnation. In: Proceedings of the 9th annual genetic and evolutionary computation conference, GECCO 2007, July 7, 2007–July 11, 2007, London, United kingdom. Association for Computing Machinery
18. Liu JH (2009) Basic theory of particle swarm algorithm and its improvement research [D]. Central South University

Chapter 5
Stability Analysis of the Standard PSO

The study of parameter selection and stability in PSO primarily relies on various stability theories of dynamic systems. Trelea [1] used a constant-coefficient linear dynamic system to determine the conditions for parameter selection and conduct stability analysis for PSO. Clerc [2] developed a constrained PSO model described by five parameters, analyzing its convergence and the characteristics of particle trajectory within the phase plane. Zeng et al. [3] proposed a continuous model of PSO and provided conditions for the asymptotic convergence of the evolution equation under different parameter settings. In Cui and Zeng [4], an improved PSO algorithm was further constructed for a system comprising integral and oscillatory components, with a discussion on methods for parameter selection. However, the dynamic equation governing particle motion in PSO is not a linear time-invariant system, which introduces certain limitations to analyses based on time-invariant dynamic systems. Kadirkamanathan [5] modeled the *gbest* particle, which falls into stagnation, as a nonlinear feedback system and analyzed the sufficient conditions for PSO stability using the *Lyapunov* stability principle. Due to its conservative nature, this approach imposes strict constraints on parameter selection, limiting particle motion to small and local oscillations. Martínez [6, 7] demonstrated that PSO can be physically modeled as a discrete stochastic damped mass-spring system and inferred that different PSO models distinct performances in balancing exploration and exploitation. The convergence of these models is linked to the stability of both first-order and second-order systems. Pan et al. [8] treated PSO as a dynamic time-varying system and analyzed the sufficient conditions for its stability without the constraint of the *Lipschitz* condition, expanding the valid range of the inertia weight ω to $(-1, 1)$. Further analysis demonstrated the algorithm's search performance when the swarm stagnates at a local extremum in the solution space [8–11], as well as its similarity to BBPSO [12].

Considering the basic PSO and combining it with Eq. (2.1), the dynamic equation of the PSO algorithm can be expressed as follows:

© Beijing Institute of Technology Press 2025
F. Pan et al., *Particle Swarm Optimizer and Multi-Objective Optimization*,
https://doi.org/10.1007/978-981-95-3381-7_5

$$x_i^{(k+1)} = \left(1 + \omega - \varphi^{(k+1)}\right) \cdot x_i^{(k)} - \omega \cdot x_i^{(k-1)} + p \qquad (5.1)$$

The dynamic model described by Eq. (5.1) has the same form as that given in Eq. (3.3). The following analysis evaluates the stability of the PSO algorithm under both time-invariant and time-varying dynamic systems, without the constraints imposed by the *Lipschitz* condition (2.2).

5.1 Dynamic System with Constant Coefficients in PSO

Equation (5.1) is a general expression. Its characteristic equation is given by:

$$D(z) = z^2 - \left(1 + \omega - \varphi^{k+1}\right) \cdot z + \omega$$

At time step k, For individual particles, the second-order discrete dynamic system described by Eq. (4.1) is analyzed using the *Jury* criterion. The conditions for all eigenvalues to be within the unit circle are depicted in Figs. 5.1 and 5.2.

$$\begin{cases} \varphi > 0 \\ 2 + 2 \cdot \omega - \varphi^{k+1} > 0 \\ |\omega| < 1 \end{cases} \qquad (5.2)$$

According to the *Jury* criterion, the distribution of eigenvalues is plotted in Fig. 5.1. *Area* 1 corresponds to the overdamped region, while *Area* 2 represents the under-damped oscillation region, with other areas indicating instability. The upper part of *Area* 2 corresponds to the left side of the unit circle in the z plane, whereas

Fig. 5.1 Distribution of eigenvalues with respect to ω and $\varphi^{(k+1)}$

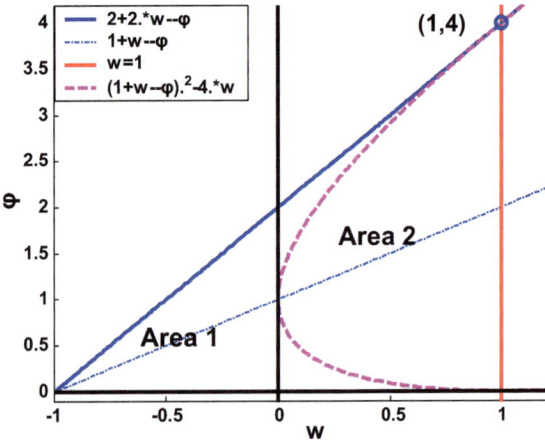

Fig. 5.2 Distribution of
eigenvalues within the unit
circle on the *z*-plane

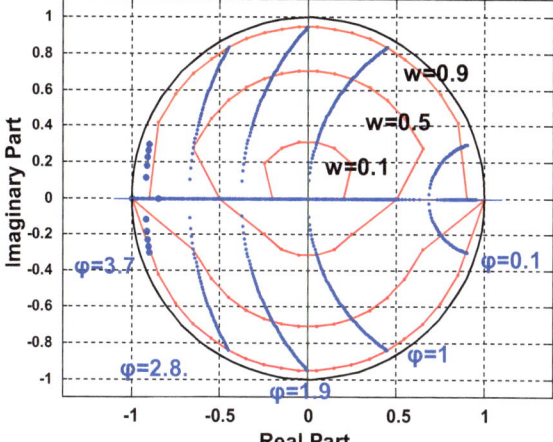

the lower part corresponds to the right side. The line $1 + \omega - \varphi^{(k+1)} = 0$ demarcates the boundary between positive and negative real parts of the eigenvalues. As ω increases, the magnitude of the eigenvalues also increases. An increase in $\varphi^{(k+1)}$ causes the eigenvalues within the unit circle to shift from right to left, decreasing the damping coefficient and resulting in more pronounced oscillations. This is consistent with the physical interpretation of the dynamic equation in PSO: when $\varphi^{(k+1)}$ is large, it indicates increased random noise, which in turn enhances particle activity.

Figure 5.2 illustrates the distribution of stable eigenvalues within the unit circle on the *z* plane. It includes the eigenvalue distributions for various values of $\varphi^{(k+1)}$ at specific values of ω, as well as the eigenvalue distributions for different values of ω at specific values of $\varphi^{(k+1)}$. Other potential values of ω and $\varphi^{(k+1)}$ are located on the real axis outside the unit circle.

The analysis of the basic PSO does not originally account for the impact of time-varying parameters in the dynamic equation on the characteristics of particle motion dynamics. Therefore, the derived conclusions can only qualitatively characterize the behavioral properties of PSO to a limited extent. In the subsequent section, we will provide a more granular analysis by taking into account the time-varying characteristics of the parameters.

5.2 Time-Varying Dynamic System in PSO

In the PSO algorithm, several critical parameters (e.g., the inertia weight, acceleration factors, and proportional constriction coefficients) significantly influence the algorithm's stability and convergence characteristics. The calibration of these parameters is predominantly empirical. *Clerc* and *Kennedy* simplified the state equation and analyzed its stability as a constant-coefficient linear system. They assumed that $p_{id}^{(k)}$ and $p_{ld}^{(k)}$ remain unchanged throughout the optimization, proving that when $\sqrt{2 \cdot \left(1 + \eta \cdot \omega - \eta \cdot \varphi^{(k)}\right)^2 - 4 \cdot \omega} < 2$ is satisfied, $x_{id}^{(k)}$ converges to point p. *Bergh* further demonstrated that the standard PSO cannot guarantee either global or local convergence.

Section 5.1 presented the distribution of eigenvalues for a system with constant coefficients. In the context of a constant-coefficient dynamic system, its motion is asymptotically stable when all eigenvalues are within the unit circle. This implies that when the acceleration factors are fixed constants, the particle trajectory will asymptotically follow the *gbest*. Consequently, all particles quickly converge to the *gbest*, potentially leading to premature convergence and preventing the algorithm from further exploration and optimization.

However, in the actual PSO algorithm, the acceleration factor φ^{k+1} for each particle varies with time. For discrete time-varying dynamic systems, it is not possible to determine asymptotic stability solely by examining whether the eigenvalues lie within the unit circle. In this section, building upon the previous analysis, we will derive the stability conditions of the dynamic system under time-varying parameters in PSO. First, we rearrange Eqs. (3.1) and (3.2) as follows:

$$\begin{bmatrix} x^{(k+1)} \\ x^{(k)} \end{bmatrix} = \begin{bmatrix} \left(1 + \eta \cdot \omega - \eta \cdot \varphi^{(k+1)}\right) & -\eta \cdot \omega \\ 1 & 0 \end{bmatrix} \cdot \begin{bmatrix} x^{(k)} \\ x^{(k-1)} \end{bmatrix} + \begin{bmatrix} \varphi^{(k+1)} \\ 0 \end{bmatrix} p \quad (5.3)$$

Equation (2.2), which serves as a velocity vector limiter during each iteration, essentially ensures that the system's bounded-input bounded-output (BIBO) stability. After transformation, it can be expressed as the *Lipschitz* condition for the dynamic system:

$$\left| x_{id}^{(k+1)} - x_{id}^{(k)} \right| \leq V_{\max} \quad (5.4)$$

In this section, we will establish sufficient conditions for the asymptotic stability of the PSO algorithm with a relaxed *Lipschitz* constraint. To begin, we introduce the conditions for asymptotic stability in a discrete dynamic system.

The state-space equation for a zero-input, time-varying discrete system can be expressed as:

$$x(k + 1) = A(k)x(k) \quad (5.5)$$

Alternatively, it can be written as:

$$x(k + 1) = T(k, 0)x(k) \quad (5.6)$$

where $x(0) = [x_1(0)x_2(0) \cdots x_N(0)]^T$, $x(k) = [x_1(k)x_2(k) \cdots x_N(k)]^T$ and

$$T(k, 0) = A(k)A(k - 1) \cdots A(0)$$

$T(k.0)$ is defined as the transfer matrix at step k.

Theorem 5.1 *If there exists a finite integer $M \geq 2$ such that for all $k \geq M$, the following condition is satisfied:*

$$\rho(T(k, k - M)) < 1$$

Then the discrete time-varying system described in Eq. (4.5) is asymptotically stable, where ρ is the spectral radius. The system's transfer matrix $T(k, k - M)$ can be obtained from Eq. (5.6):

$$T(k, k - M) = \prod_{i=1}^{M} A(k - i)$$

Lemma 5.1 *A sufficient condition for the asymptotic stability of the dynamic system in PSO is that the acceleration factor $\varphi^{(k+1)}$ satisfies the following conditions:*

$$\eta \cdot \omega \in (-1, 1) \tag{5.7}$$

$$\left\{ \eta \cdot \varphi^{(k+1)} \in U | U \subset \left(\varphi_{min}^{(k+1)}, \varphi_{max}^{(k+1)} \right) \right\} \tag{5.8}$$

$$\varphi_{min}^{(k+1)} = \frac{-\eta \varphi^{(k)}(1 + \eta \omega)}{1 + \eta \omega - \eta \varphi^{(k)}}$$

$$\varphi_{max}^{(k+1)} = \frac{2\left(1 + (\eta \omega)^2\right) - \eta \varphi^{(k)}(1 + \eta \omega)}{1 + \eta \omega - \eta \varphi^{(k)}}$$

Proof By rearranging the equation of standard PSO, we get:

$$\begin{bmatrix} x_{id}^{(k+1)} \\ x_{id}^{(k)} \end{bmatrix} = \begin{bmatrix} \left(1 + \eta \cdot \omega - \eta \cdot \varphi^{(k+1)}\right) & -\eta \cdot \omega \\ 1 & 0 \end{bmatrix} \cdot \begin{bmatrix} x_{id}^{(k)} \\ x_{id}^{(k-1)} \end{bmatrix} + \begin{bmatrix} \varphi^{(k+1)} \\ 0 \end{bmatrix} p \tag{5.9}$$

where $\varphi_i^{(k+1)} = c_i \times r_i$, $\varphi^{(k+1)} = \varphi_1^{(k+1)} + \varphi_2^{(k+1)}$, $p = \left(\varphi_1^{(k+1)} \cdot p_{id}^{(k)} + \varphi_2^{(k+1)} \cdot p_{ld}^{(k)}\right) / \left(\varphi_1^{(k+1)} + \varphi_2^{(k+1)}\right)$, and η is the proportional constriction coefficient, and p is the system's external input. Thus, the transfer matrix of Eq. (5.9) is:

$$T(k, k - 2) = \begin{bmatrix} 1 + \eta \cdot \omega - \eta \cdot \varphi^{(k)} & -\eta \cdot \omega \\ 1 & 0 \end{bmatrix} \cdot \begin{bmatrix} 1 + \eta \cdot \omega - \eta \cdot \varphi^{(k+1)} & -\eta \cdot \omega \\ 1 & 0 \end{bmatrix} \tag{5.10}$$

According to Theorem 5.1, when the condition $\rho(T(k, k - 2)) = \max \|T(k, k - 2)\| = \max \lambda_i < 1$ is satisfied, the dynamic system described by

Eq. (5.9) is asymptotically stable.

$$|\lambda E - T(k, k-2)| = \begin{vmatrix} \lambda - L^{(k)} L^{(k+1)} + \eta \cdot \omega & \eta \cdot \omega \cdot L^{(k)} \\ -L^{(k)} & \lambda + \eta \cdot \omega \end{vmatrix}$$

$$L^{(k)} = \left(1 + \eta \cdot \omega - \eta \cdot \varphi^{(k)}\right)$$

The characteristic equation of the transfer matrix is expressed as follows:

$$D(\lambda) = \lambda^2 + \lambda \cdot \left[2\eta\omega - L^{(k)}(1 + \eta\omega) + L^{(k)} \cdot \eta \cdot \varphi^{(k+1)}\right] + (\eta\omega)^2 \qquad (5.11)$$

According to the *Jury* criterion, a necessary and sufficient condition for the spectral radius $\rho(\cdot) < 1$ in Eqs. (5.11) and (4.11) is:

$$(\eta\omega)^2 < 1 \qquad (5.12)$$

$$\left|2\eta\omega - L^{(k)}(1 + \eta\omega) + L^{(k)} \cdot \eta \cdot \varphi^{(k+1)}\right| < 1 + (\eta\omega)^2 \qquad (5.13)$$

Solving Eq. (5.12) yields:

$$\eta \cdot \omega \in (-1, 1)$$

Solving Eq. (5.13) yields:

$$\frac{-\eta\varphi^{(k)}(1 + \eta\omega)}{1 + \eta\omega - \eta\varphi^{(k)}} \leq \eta \cdot \varphi^{(k)} \leq \frac{2\left(1 + (\eta\omega)^2\right) - \eta\varphi^{(k)}(1 + \eta\omega)}{1 + \eta\omega - \eta\varphi^{(k)}}$$

This implies that, at any given time step $k+1$, the value ranges of the acceleration factor $\eta \cdot \varphi^{(k+1)}$ and the inertia weight ω are defined by Eqs. (2.3) and (2.4). When the range $\eta \cdot \varphi^{(k+1)} \in U$ satisfies $U \subset \left(\varphi_{min}^{(k+1)}, \varphi_{max}^{(k+1)}\right)$, the system is asymptotically stable. If this range is exceeded, however, the system tends to diverge. Figures 5.3 and 5.4 show the upper and lower boundary surfaces of the stability region for $\eta \cdot \varphi^{(k+1)}$ when w ranges between $(-1, 1)$ and $\eta \cdot \varphi^{(k)}$ lies between $(0, 4)$. Figure 5.5 illustrates the variation range of $\eta \cdot \varphi^{(k+1)}$, that is, $\varphi_{max}^{(k+1)} - \varphi_{min}^{(k+1)}$.

Corollary 5.1 *When the change rate $\eta \cdot \Delta\varphi^{(k)}$ of the acceleration factor $\varphi^{(k+1)}$ meets the following conditions:*

$$\eta \cdot \omega \in (-1, 1)$$

$$\left\{\eta \cdot \Delta\varphi^{(k)} \in U \mid U \subset [\Delta_{min}, \Delta_{max}]\right\}$$

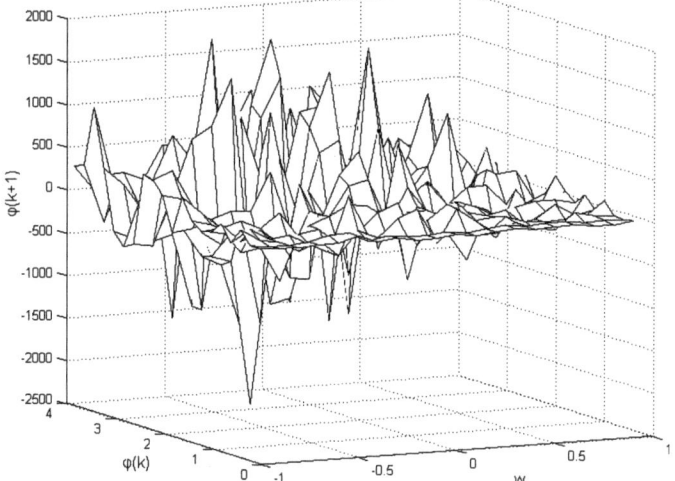

Fig. 5.3 Surface of the value range for $\varphi_{\max}^{(k+1)}$

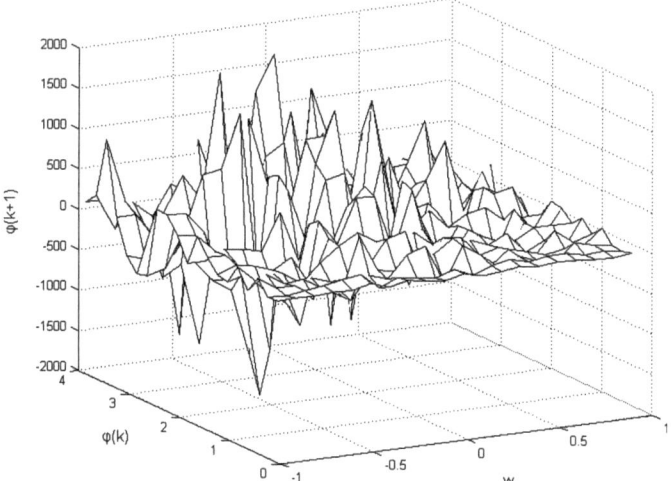

Fig. 5.4 Surface of the value range for $\varphi_{\min}^{(k+1)}$

$$\Delta_{\min} = \frac{\left(1 + \eta\omega - \eta\varphi^{(k)}\right)^2 - (1 + \eta\omega)^2}{1 + \eta\omega - \eta\varphi^{(k)}}$$

The dynamic system of PSO is asymptotically stable.

Proof Since

$$\varphi^{(k+1)} = \varphi^{(k)} + \eta \cdot \Delta\varphi^{(k)} \tag{5.14}$$

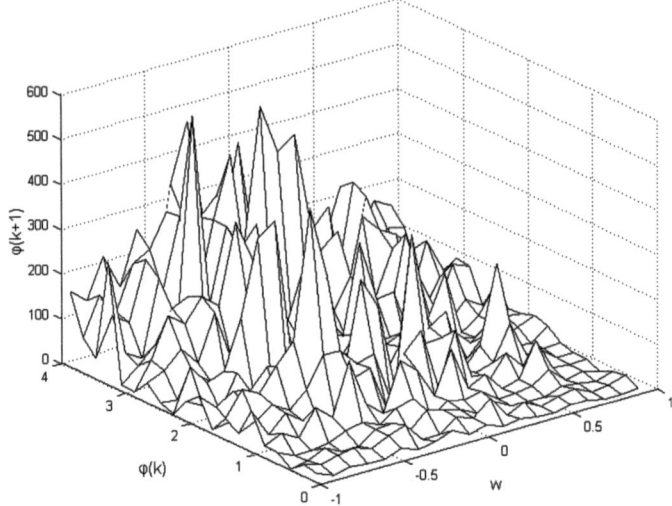

Fig. 5.5 Variation range of $\eta \cdot \varphi^{(k+1)}$: $\varphi_{max}^{(k+1)} - \varphi_{min}^{(k+1)}$

From Lemma 5.1, the condition for the stability of the acceleration factor is:

$$\frac{-\eta\varphi^{(k)}(1 + \eta\omega)}{1 + \eta\omega - \eta\varphi^{(k)}} \leq \eta \cdot \varphi^{(k)} \leq \frac{2\left(1 + (\eta\omega)^2\right) - \eta\varphi^{(k)}(1 + \eta\omega)}{1 + \eta\omega - \eta\varphi^{(k)}}$$

By subtracting $\varphi^{(k)}$ from both sides of the equation, we obtain (Fig. 5.6):

$$\frac{\left(1 + \eta\omega - \eta\varphi^{(k)}\right)^2 - (1 + \eta\omega)^2}{1 + \eta\omega - \eta\varphi^{(k)}} \leq \Delta\eta \cdot \varphi^{(k)} \leq \frac{\left(1 + \eta\omega - \eta\varphi^{(k)}\right)^2 + (1 - \eta\omega)^2}{1 + \eta\omega - \eta\varphi^{(k)}}$$

$$(5.15)$$

5.3 Validation Experiment

Lemma 5.1 provides the range of the acceleration factor $\varphi^{(k+1)}$ and the inertia weight w for the stability of the PSO algorithm. Corollary 5.1 outlines the conditions under which the acceleration factor ensures that the dynamic system in PSO remains asymptotically stable. Based on this theoretical foundation, this section introduces two improved algorithms: the inertia factor harmonious particle swarm optimization (IFHPSO) and the acceleration factor increment harmonious particle swarm optimization (AFIHPSO). Extensive experiments were conducted to validate the correctness of this theory.

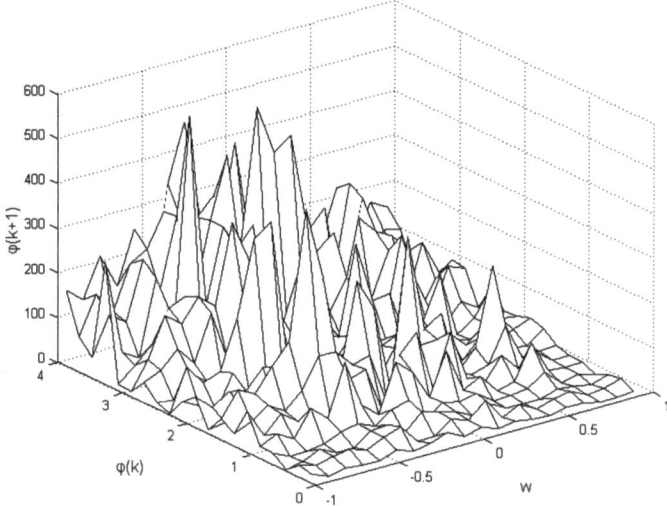

Fig. 5.6 Variation range of $\Delta \eta \cdot \varphi^{(k)}$: $\Delta_{\max} - \Delta_{\min}$

5.3.1 Inertia Weight Harmonious Particle Swarm Optimization (IFHPSO)

5.3.1.1 Algorithm Description

As analyzed in Chap. 3, when the swarm diversity decreases significantly, the swarm may stagnate at a local optimum, losing its ability to explore new areas. However, the swarm should not always maintain a high level of diversity; that is, it should not always perform global searches over a wide range while neglecting more refined local searches. In the standard PSO, parameter selection is mostly based on empirical rules, which cannot guarantee optimal performance. Therefore, based on the stability analysis from Chap. 4, this section proposes an improved method in which the values of acceleration factor meet the stability conditions outlined in Eq. (5.8) in Sect. 5.2. The swarm's inertia weight ω is adjusted according to Eq. (5.7) to maintain a balance in swarm diversity. Additionally, as discussed in Sect. 3.1, the star-shaped information transmission topology is a primary factor contributing to premature convergence. To address this, a crossover operator is introduced into the swarm. Although the use of crossover operators is generally not encouraged in evolutionary computation, it is incorporated here to enhance information exchange between particles. The diversity of the swarm is measured using Eq. (3.4).

Equation (5.7) defines the range for the stable inertia weight ω in the PSO algorithm. In order to ensure stability, ω must lie between $(-1, 1)$. If ω falls outside this range, the algorithm will inevitably become unstable. Figures 5.1 and 5.2 in Sect. 5.1 depict the distribution of eigenvalues for the dynamic system at different values of ω and φ^{k+1}, providing a qualitative description of how these parameter values influence the algorithm. As shown in Fig. 5.2, a smaller ω results in the eigenvalue distribution on the Z-plane moving closer to the center of the unit circle. Similarly, a smaller φ brings the eigenvalues closer to the real axis. Figure 5.2 also demonstrates that as both ω and φ^{k+1} decrease, the dynamic equation occupies a larger portion of the overdamped region, thereby accelerating the search process. This is consistent with the physical interpretation: the smaller the ω, the faster the velocity decays, and the smaller the φ^{k+1}, the weaker the external disturbances on the particles, leading to a focus on local search. The parameters of the PSO algorithm can be chosen based on the following logic: both ω and φ^{k+1} should be set to smaller values to enhance the search and contraction capabilities. When the swarm density becomes too high and expansion is desired, ω can be selected outside the range of $(-1, 1)$, as illustrated in Fig. 5.7.

$$if\ (Div < dlow)\&(ft > CF_{max})$$
$$\omega = -1$$
$$else\ if\ (Div > dhigh)$$
$$\omega = 0.1$$
$$else$$
$$\omega = default$$

Where Div represents a measure of swarm diversity; ft represents the number of consecutive iterations without improvement in fitness; $dhigh$ represents the upper limit of Div; $dlow$ represents the lower limit of Div; CF_{max} represents the upper limit of ft.

The crossover operator used in the algorithm is defined as follows:

$$x_i^{k+1} = \left[x_{i1}^k \cdots x_{ip}^k, x_{jp+1}^k \cdots x_{jn}^k \right]$$
$$x_j^{k+1} = \left[x_{j1}^k \cdots x_{jp}^k, x_{ip+1}^k \cdots x_{in}^k \right]$$

p is a crossover point randomly selected, and the crossover operator is applied to both the position and velocity vectors of the two particles from the previous time step, generating two offspring for the next iteration. The introduction of the crossover operator expands the pathways for information exchange within the swarm. As depicted

Fig. 5.7 Selection criteria for the inertia weight ω

$$if\ (Div < dlow)\ \&\ (ft > CF_{max})$$
$$\omega = -1$$
$$else\ if\ (Div > dhigh)$$
$$\omega = 0.1$$
$$else$$
$$\omega = default$$

Fig. 5.8 Pseudo-star
topology for information
transmission after
introducing the crossover
operator

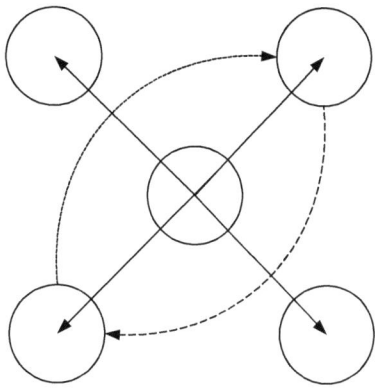

in Fig. 5.8, the dashed lines represent temporary information exchange channels, while the arrows signify bidirectional information flow. The topology of the swarm is no longer a strictly star structure. Moreover, these temporary information channels, established through crossover, do not exist in every cycle, nor are they fixed between specific pairs of individuals. Therefore, the topology under this model is referred to as a *"pseudo-star topology."*

Based on the above improvements, the steps for IFHPSO are outlined as follows:

Step 1: Initialize the positions x^{k_0}, initial velocities v^{k_0}, *gbest* solution $p_g^{k_0}$, and *pbest* solution $p_i^{k_0}$;

Step 2: Determine the inertia weight ω based on the selection criteria and set its value;

Step 3: Randomly select two particles to perform the crossover operation;

Step 4: Update the states of the remaining particles in the swarm using Eqs. (2.1) and (2.2);

Step 5: Update the *gbest* and *pbest* solutions, then calculate *Div* and *ft*;

Step 6: If the termination condition is met, stop the process; otherwise, return to Step 2.

5.3.1.2 Results of the Simulation Experiment

① Experiment design and parameter settings.

In the simulation experiments, the standard PSO, standard GA, and IFHPSO were employed to optimize four *Benchmark* functions: *Sphere, Rosenbrock, Griewank, and Rosenbrock* (see Appendix). Two sets of optimization simulations were conducted:

(1) Over a fixed period, 50 optimization trials were performed for each of the four functions across different dimensions. The statistical results of the three

Table 5.1 Search space and initialization ranges for *Benchmark* functions

Functions	Search space	Initialization range
Sphere	$[-100, 100]$	$[-50, 100]$
Rosenbrock	$[-100, 100]$	$[-15, 30]$
Griewank	$[-600, 600]$	$[-300, 600]$
Rastrigin	$[-10, 10]$	$[-5, 10]$

algorithms are shown in Tables 5.2, 5.4 and 5.5 (values less than 10^{-10} are considered zero) (Tables 5.1 and 5.3).

2 Optimization was carried out on the 50-dimensional *Rosenbrock* and *Rastrigin* functions until the IFHPSO results met the required standards. Figures 5.9 and 5.10 compare the fitness and diversity of PSO and IFHPSO under identical conditions and timeframes.

Table 5.2 Results of 50 optimization trials for different dimensions of the *Sphere* function using standard PSO, GA, and IFHPSO

f_1		*Sphere*			
Dimension	Period	Fitness	PSO ($\times 10^{-10}$)	GA ($\times 10^{-8}$)	IFHPSO ($\times 10^{-10}$)
2	500	Mean	29.028	2.78937	0
		Variance	57,290	5.59642	0
10	1000	Mean	58.537	388.21	0
		Variance	12.977	638.68	0
20	2000	Mean	4.9526	2955.9	8.6272
		Variance	14,809	2771.1	7.0972
30	3000	Mean	8.0006	73,693	36.003
		Variance	18,499	43,453	34.641

Table 5.3 Results of 50 optimization trials for different dimensions of the *Rosenbrock* function using standard PSO, GA, and IFHPSO

f_2		*Rosenbrock*			
Dimension	Period	Fitness	PSO	GA	IFHPSO
2	500	Mean	0.284	5.741	1.777×10^{-5}
		Variance	0.217	9.547	2.685×10^{-5}
10	1000	Mean	5.1958	87.739	3.333
		Variance	3.4202	9.353	1.501
20	2000	Mean	12.083	136.748	8.445
		Variance	6.193	19.222	5.150
30	3000	Mean	78.418	180.28	14.3413
		Variance	14.809	25.45	9.6067

Table 5.4   Results of 50 optimization trials for different dimensions of the *Griewank* function using standard PSO, GA, and IFHPSO

f_3		*Griewank*			
Dimension	Period	Fitness	PSO ($\times 10^{-3}$)	GA	IFHPSO ($\times 10^{-3}$)
2	200	Mean	3.915	1.095	4.807
		Variance	3.1053	0.554	3.619
10	1000	Mean	64.611	1.472	60.875
		Variance	31.015	0.431	33.689
20	2000	Mean	27.332	2.256	34.151
		Variance	23.147	1.952	27.300
30	3000	Mean	5.475	2.761	3.0905
		Variance	9.035	1.9326	3.3385

Table 5.5   Results of 50 optimization trials for different dimensions of the *Rastrigin* function using standard PSO, GA, and IFHPSO

f_4		*Rastrigin*			
Dimension	Period	Fitness	PSO	GA	IFHPSO
2	200	Mean	0.0845	0.167	0
		Variance	0.0671	0.183	0
10	1000	Mean	3.1061	6.719	1.486×10^{-8}
		Variance	1.118	1.189	3.389×10^{-8}
20	2000	Mean	16.981	18.83	1.664
		Variance	4.464	8.660	1.179
30	3000	Mean	40.019	37.83	11.297
		Variance	7.387	7.512	6.025

Fig. 5.9 Fitness and diversity of IFHPSO and PSO in optimizing the 50-dimensional *Rosenbrock* function

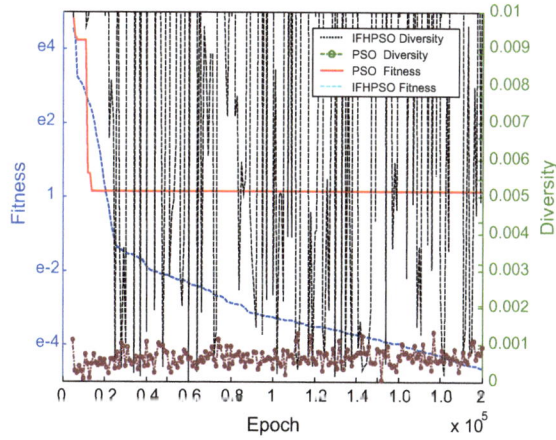

Fig. 5.10 Fitness and
diversity of IFHPSO and
PSO in optimizing the
50-dimensional *Rastrigin*
function

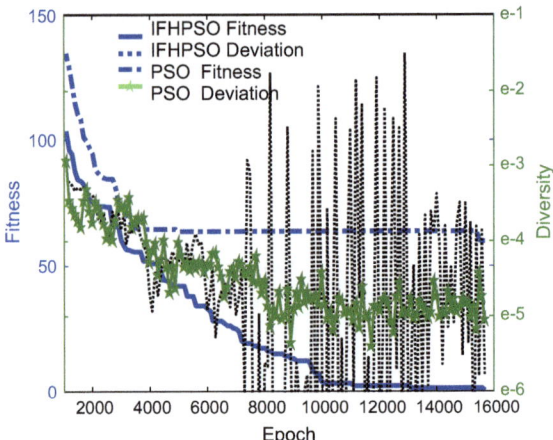

For all three algorithms, the swarm size was set to 30. In both PSO and IFHPSO, the parameters c_1 and c_2 were configured to 2. The inertia weight ω was a variable that decreased from 0.6 to 0.1 as the fitness improved. The maximum velocity vector V_{max} was set to half the distance of the search space. The diversity indicators *dlow* and *dhigh* were set to 10^{-3} and 0.25, respectively, with the failure threshold f_c set to 20. Table 5.1 shows the search space and initialization ranges for the *Benchmark* functions.

② Experimental results and analysis.

When optimizing the multimodal *Rastrigin* function, PSO performance declines as dimensionality increases due to the proliferation of local extrema. The *Griewank* function, which can be seen as a *Sphere* function added with the noise term $\prod \cos\left((x_i - 1)/\sqrt{i}\right)$, becomes smoother as the dimensionality increases since the noise term tends towards 0. Therefore, both PSO and IFHPSO exhibit robust optimization capabilities across both low and high dimensions. This highlights the effective local optimization capability of PSO in search spaces with fewer local extrema. In contrast, IFHPSO consistently demonstrates robust optimization performance across different scenarios. Figures 5.9 and 5.10 depict comparative analyses of the fitness and diversity of IFHPSO and PSO when optimizing the 50-dimensional *Rosenbrock* and *Rastrigin* functions. The optimization for IFHPSO was terminated once the fitness and target values met the specified criteria (i.e., the error was less than 10^{-3}). As illustrated, PSO tends to fall into premature convergence as diversity decreases, whereas IFHPSO maintains fitness and effectively balances swarm contraction and expansion, ensuring continuous convergence throughout the optimization process.

5.3.2   Acceleration Factor Increment Harmonious Particle Swarm Optimization (AFHPSO)

In Sect. 5.3.1, an improved algorithm was proposed, which adjusts the velocity inertia weight ω to coordinate the search with PSO, as discussed in Lemma 5.1 of Chap. 4. Simulation results confirmed its effectiveness. This section introduces the AFIHPSO algorithm, which builds on the conclusions of Lemma 5.1 and Corollary 5.1.

5.3.2.1   Description of AFHPSO

At each time step, the acceleration factor $\eta \cdot \varphi^{(k)}$ can be dynamically adjusted to determine the value of $\eta \cdot \varphi^{(k+1)}$ for the next period: $\eta\varphi^{(k+1)} = \varphi(\eta\varphi^{(k)}, \omega)$. When ω satisfies the condition given in Eq. (5.7), whether $\eta \cdot \varphi^{(k+1)}$ meets the condition in Eq. (5.8) determines the stability of the particle's individual movement, which directly affects whether the swarm contracts or expands. Thus, under certain conditions, by adjusting these parameters, the overall behavior of the swarm can be controlled, allowing for continuous cycles of expansion and contraction throughout the search process, thereby facilitating repeated exploration of the solution space.

Therefore, when swarm divergence is required, $\eta \cdot \varphi^{(k+1)}$ is determined using the following equation:

$$\eta \cdot \varphi^{(k+1)} = \left(\varphi_{max}^{(k+1)} + \varphi_{min}^{(k+1)}\right) + \frac{abs\left(\varphi_{max}^{(k+1)} - \varphi_{min}^{(k+1)}\right)}{2} \cdot randn \qquad (5.16)$$

When swarm convergence is required, $\eta \cdot \varphi^{(k+1)}$ is derived using the following equation:

$$\eta \cdot \varphi^{(k+1)} = \frac{\left(\varphi_{max}^{(k+1)} + \varphi_{min}^{(k+1)}\right)}{2} + \frac{abs\left(\varphi_{max}^{(k+1)} - \varphi_{min}^{(k+1)}\right)}{4} \cdot randn \qquad (5.17)$$

where $\varphi_{max}^{(k+1)}$ and $\varphi_{min}^{(k+1)}$ are defined by Eq. (5.8), and *randn* represents a random number between $(-1, 1)$. The steps for implementing the AFHPSO algorithm are as follows:

Step 1: Initialize the swarm's initial positions $x^{(k_0)}$, initial velocities $v^{(k_0)}$, *gbest* solution $p_g^{(k_0)}$, and *pbest* solutions $p_i^{(k_0)}$;

Step 2: Check if the termination condition is met. If so, stop the operation and exit; otherwise, proceed to the next iteration;

Step 3: If $Div \geq DV_{max}$, calculate $\eta \cdot \varphi^{(k+1)}$ using Eq. (5.17) to contract the swarm;

Step 4: If $Div \leq DV_{min}$ and $ft > CF_{max}$, calculate $\eta \cdot \varphi^{(k+1)}$ using Eq. (5.16) to expand the swarm.

Step 5: Otherwise, update the states of the remaining particles in the swarm using Eqs. (2.1) and (2.2);

Step 6: Update the *gbest* and *pbest* solutions, then calculate *Div* and *ft*. Return to Step 2.

5.3.2.2 Acceleration Factor Increment Harmonious Particle Swarm Optimization (AFIHPSO)

Based on Corollary 5.1 and the AFHPSO algorithm discussed in Sect. 5.3.2.1, another form of the improved harmonious PSO algorithm, known as AFIHPSO, can be derived. AFHPSO directly obtains the value of $\eta \cdot \varphi^{(k+1)}$ for the next iteration based on its range obtained from the previous parameter conditions. In contrast, AFIHPSO determines the variation range of the value near $\eta \cdot \varphi^{(k)}$ under the previous parameter conditions.

When the swarm needs to expand, the value of $\eta \cdot \varphi^{(k+1)}$ is formulated using the following equation:

$$\eta \cdot \varphi^{(k+1)} = \eta \cdot \varphi^{(k)} + abs(\Delta_{max} - \Delta_{min}) \cdot randn \tag{5.18}$$

When the swarm needs to contract, the value of $\eta \cdot \varphi^{(k+1)}$ is calculated using the following equation:

$$\eta \cdot \varphi^{(k+1)} = \eta \cdot \varphi^{(k)} + \frac{abs(\Delta_{max} - \Delta_{min})}{4} \cdot randn \tag{5.19}$$

where Δ_{max}, Δ_{min}, and $\varphi_{min}^{(k+1)}$ are defined by Corollary 5.1. The description of AFIHPSO is as follows:

Step 1: Initialize the swarm's initial positions $x^{(k_0)}$, initial velocities $v^{(k_0)}$, *gbest* solution $p_g^{(k_0)}$ and *pbest* solutions $p_i^{(k_0)}$;

Step 2: Check if the termination condition is met. If so, stop the operation and exit; otherwise, proceed to the next iteration;

Step 3: If $Div \geq DV_{max}$, calculate $\eta \cdot \varphi^{(k+1)}$ using Eq. (5.19) to contract the swarm;

Step 4: If $Div \leq DV_{min}$ and $ft > CF_{max}$, calculate $\eta \cdot \varphi^{(k+1)}$ using Eq. (5.18) to expand the swarm.

Step 5: Otherwise, update the states of the remaining particles in the swarm using Eqs. (2.1) and (2.2);

Step 6: Update the *gbest* and *pbest* solutions, then calculate *Div* and *ft*. Return to Step 2.

5.3.2.3 Results of the Simulation Experiment

This section employs the two harmonious PSO algorithms proposed in this chapter: AFHPSO and AFIHPSO, to optimize two distinct sets of *Benchmark* functions and verify the effectiveness of the algorithms.

In the

① Simulation results of AFHPSO.

simulation experiments, both the standard PSO and AFHPSO are used to conduct two sets of optimization experiments on four *Benchmark* functions, as shown in Table. 5.6. The *DeJong* and *Griewank* functions are simulated 50 times, with each simulation running for 3,000 optimization cycles. The statistical characteristics of these 50 simulation results are then compared. In the second experiment, the *Rastrigin* and *Rosenbrock* functions are optimized until the target is achieved.

For the parameter selection of standard PSO, the recommended parameters are $c_1 = c_2 = 1.494$, $rand_{id}$, $rand_{pd} \in rand(0, 1)$, $\omega = 0.729$, and $V_{max} = 10$. In the first experiment, the parameters for AFHPSO are chosen as $dlow = 1$, $dhigh = 10^{-5}$, and $\omega = 0.5$. In the second experiment, the parameters for AFHPSO are $dlow = 1$ and $dhigh = 10^{-3}$. According to Lemma 5.1, the range of ω is $(-1, 1)$; here, it's taken as $\omega = \{-0.5, 0.5\}$.

The results of the first experiment are shown in Table 5.7. The *Sphere* function is the most commonly used function for comparing methods of evolutionary computation. It can be used to test local optimization capabilities. Since AFHPSO adjusts the swarm diversity by expanding or contracting based on its measurement, the standard PSO converges faster than AFHPSO in the optimization of the *Sphere* function. As described in Sect. 5.3.1.2, the *Griewank* function is equivalent to adding a noise term $\prod \cos\left((x_i - 1)/\sqrt{i}\right)$ to the *Sphere* function. Since the function includes product terms between variables, its parameters are strongly correlated. As can be seen from the table, AFHPSO outperforms PSO in this case.

The results of the second experiment are presented in Figs. 5.11 and 5.12. A comparison of the fitness between AFHPSO and PSO for optimizing the 30-dimensional *Rosenbrock* and *Rastrigin* functions is presented. Each figure uses the point at which AFHPSO meets the optimization criteria as the termination condition

Table 5.6 Simulation experiment with *Benchmark* functions of AFHPSO

Function	Search space	Initialization range	Global optimal value	Expected error
Sphere	$[-100, 100]$	$[-50, 100]$	0	0
Griewank	$[-600, 600]$	$[-300, 600]$	0	0
Rosenbrock	$[-100, 100]$	$[-15, 30]$	0	10^{-3}
Rastrigin	$[-10, 10]$	$[-5, 10]$	0	10^{-3}

Table 5.7 Simulation results of 50 optimizations on 30-dimensional *Sphere* and *Griewank* functions using Standard PSO and AFHPSO

Algorithm	Function	Fitness mean value	Fitness variance value
Standard*PSO*	*DeJong*	0	0
	Griewank	1.487×10^{-10}	$8,068 \times 10^{-10}$
AFHPSO	*DeJong*	5.982×10^{-9}	6.202×10^{-9}
	Griewank	$9,569 \times 10^{-11}$	1.843×10^{-11}

(defined as when the error between the optimized fitness value and the target value is less than 10^{-3}).

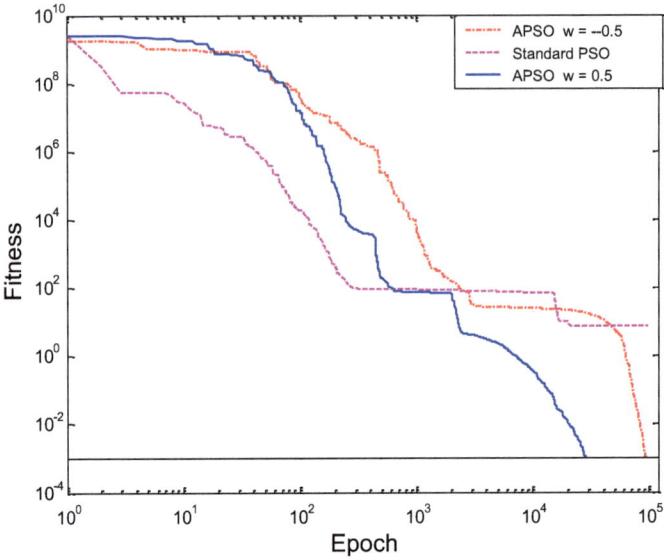

Fig. 5.11 Optimization results of the 30-dimensional *Rosenbrock* function using AFHPSO and PSO

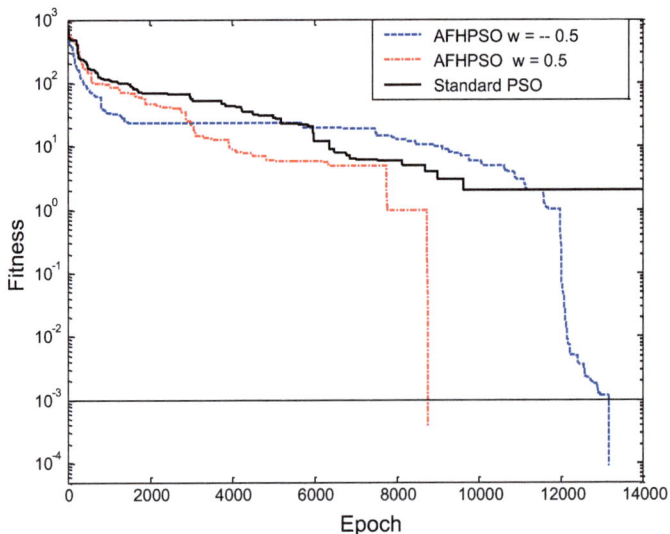

Fig. 5.12 Optimization results of the 30-dimensional *Rastrigin* function using AFHPSO and PSO

As shown in the figures, PSO quickly stagnates as diversity decreases, leading to premature convergence. In contrast, AFHPSO continuously adjusts the diversity, ensuring sustained convergence throughout the entire optimization process. When $\omega = 0.5$, the eigenvalues are distributed across both the overdamped and underdamped regions. However, when $\omega = -0.5$, all eigenvalues fall within the overdamped region, resulting in strong convergence of particle motion but weakening the algorithm's ability to explore a broader search space. This is why the optimization cycle shown in the results is longer when $\omega = 0.5$.

② Simulation results of AFIHPSO.

The AFIHPSO and standard PSO were used to perform optimization simulations on four *Benchmark* functions.

The parameter selection follows the standard PSO settings: $c_1 = c_2 = 2$, $rand_{id}$, $rand_{pd} \in rand(0, 1)$. Additionally, based on Eq. (5.2), the proportional constriction factor η has an inverse effect on the ranges of ω, φ, and $\Delta\varphi$, either expanding or contracting their values. Here, η is set to 1. Neither method imposes the *Lipschitz* condition with a V_{max} constraint. The swarm size is set to 30.

The three functions listed in Table 5.8 were optimized within 3,000 iterations. The termination condition was either reaching the target optimization error or the maximum number of iterations. Each parameter setting was simulated 50 times, and the statistical characteristics of these 50 simulation results were compared. The diversity metric, defined based on swarm variance, was used to measure the convergence, as specified in Eq. (3.4). The results are presented in Tables 5.9, 5.10, 5.11, 5.12, 5.13 and 5.14, showing the simulation outcomes for both methods when the inertia weight ω is within the range $(-1, 1)$.

First, we analyze the standard PSO. Without the constraints of the *Lipschitz* condition, the statistical characteristics of the optimization are presented in Tables 5.10 and 5.12. The data indicate that, under six different values of ω, the algorithm retains its search capability only when ω is around 0.5. In other cases, it fails to achieve the optimization goal (Fig. 5.14). The reason can be inferred from the diversity metrics: when the maximum iterations is reached, the swarm diverges ($Div \to \infty$) when $\omega = 0.9$. In contrast, at $\omega = 0.1$, the swarm converges too rapidly, leading to premature convergence ($Div \to 0$). When $\omega < 0$, the entire swarm tends to diverge, and the closer ω is to -1, the faster the divergence occurs. In Table. 5.14, the optimization results for $\omega = [-0.1, 0.1, 0.5, 0.9]$ roughly meet the target. However, the mean diversity suggests that the properties of swarm divergence and overly rapid convergence are consistent with the analyses in Tables 5.10 and 5.12. The reason the

Table 5.8 Simulation experiment with *Benchmark* functions of AFIHPSO

Function	Search space	Dimension	Global optimal value	Expected error
Sphere	$[-100, 100]$	30	0	$\times 10^{-8}$
Griewank	$[-100, 100]$	30	0	$\times 10^{-5}$
Schaffer's f6	$[-100, 100]$	2	-1	$\times 10^{-3}$

Table 5.9 AFIHPSO optimization of the 30-dimensional *DeJong* function

DeJong	Fitness ($\times 10^{-7}$)				Diversity mean ($\times 10^{-9}$)	Period mean value
Inertia factor ω	Mean	Variance	Optimal value	Worst value		
0.9	7.1694	38.7063	1.8761	183.00	2905.53	2983.1
0.5	1.04501	0.475008	22.058	2.5318	9.1014	673.64
0.1	4.18171	2.62253	4.1817	12.858	9.3247	1473.66
−0.1	4.94899	3.25922	1.0886	12.895	9.3070	1030.61
−0.5	2.27601	0.566314	1.4723	3.10349	9.3647	1751.85
−0.9	41.5594	39.7923	4.8563	173.35	786.454	3000

Schaffer's f6 function can reach the optimization target is due to its low dimensionality. Shi [13, 14] provided simulation results for $V_{max} = [2, 3, 4, 5, 10, X_{max}]$. As the constraint range expands, the ω value for achieving favorable statistical characteristics regarding optimization gradually decreases. However, Shi [13, 14] did not conduct experiments or analyses under unconstrained conditions. The conclusions drawn from Tables 5.10, 5.12 and 5.14 are consistent with *Shi*'s inference.

Next, let's analyze the optimization results of AFIHPSO given in Tables 5.9, 5.11, and 5.13. Overall, the optimization of the three functions meets the target. Furthermore, after reaching the target, the mean fitness of the swarm remains within a relatively balanced range. As shown in Fig. 5.13, for each of the six simulations of each function, the performance varies significantly with different values of ω. When ω is near the boundaries of the range, the mean optimization cycle is longer than for other values. The mean diversity in Table 5.10 ($\omega = [−0.9, 0.9]$) and Table 5.13 ($\omega = 0.9$) similarly highlight these differences.

A cross-comparison of the two methods using the six tables and two figures shows that within the empirical range of $0 < \omega < 1$, the performance of the standard PSO

Table 5.10 Standard PSO optimization of the 30-dimensional *DeJong* function

DeJong	Fitness				Diversity mean	Period mean value
Inertia factor ω	Mean	Variance	Optimal value	Worst value		
0.9	15,094.1	1490.22	11,889.8	17,697.1	3.396×10^{63}	3000
0.5	9.505×10^{-9}	4.709×10^{-10}	8.394×10^{-9}	9.999009	4.219×10^{-7}	913.5
0.1	1.546	2.884	0.0144	10.833	6.298×10^{-12}	3000
−0.1	10,689.4	3075.74	5379.99	15,639.4	4.034×10^{15}	3000
−0.5	NaN	NaN	NaN	NaN	NaN	3000
−0.9	NaN	NaN	NaN	NaN	NaN	3000

Table 5.11 AFIHPSO optimization of the 30-dimensional *Griewank* function

Griewank Inertia factor ω	Fitness Mean	Variance	Optimal value	Worst value	Diversity mean × 10^{-2}	Period mean value
0.9	2.016×10^{-5}	2.662×10^{-5}	9.569×10^{-8}	1.098×10^{-4}	2.126	2730.48
0.5	3.611×10^{-6}	7.958×10^{-6}	7.958×10^{-6}	2.500×10^{-5}	3.140	2532.84
0.1	3.57×10^{-3}	7.834×10^{-3}	1.126×10^{-8}	0.0474	1.024	2431.52
−0.1	2.73×10^{-3}	3.677×10^{-3}	1.613×10^{-10}	0.0145	1.122	2196.53
−0.5	1.552×10^{-5}	2.179×10^{-5}	5.224×10^{-11}	5.831×10^{-5}	1.875	2147.5
−0.9	4.425×10^{-5}	7.538×10^{-5}	7.015×10^{-7}	3.031×10^{-4}	1.560	2757.25

Table 5.12 Standard PSO optimization of the 30-dimensional *Griewank* function

Griewank Inertia factor ω	Fitness Mean	Variance	Optimal value	Worst value	Diversity mean	Period mean value
0.9	4.727	0.484	3.1102	5.463	2.891×10^{65}	3000
0.5	0.0167	0.0137	9.040×10^{-9}	0.0394	8.904×10^{-7}	2454.7
0.1	0.0559	0.0370	9.857×10^{-3}	0.140	2.724×10^{-11}	3000
−0.1	3.593	1.005	1.810	5.100	2.252×10^{17}	3000
−0.5	NaN	NaN	NaN	NaN	NaN	3000
−0.9	NaN	NaN	NaN	NaN	NaN	3000

significantly declines. However, the PSO with the adaptive acceleration factor, as determined by Corollary 5.1, outperforms it. Few studies have explored the characteristics of standard PSO in the range of $-1 < \omega < 0$. As seen in Fig. 5.14, within this range, the distribution of eigenvalues within the unit circle is minimal, making it challenging to ensure algorithm characteristics. The simulation results also indicate that the standard PSO performs poorly in this range. However, theoretically, stable convergence is possible when $-1 < \omega < 0$. Therefore, by appropriately selecting the range of the acceleration factor, the algorithm's stability and convergence can be ensured, and the optimization results are quite satisfactory.

Table 5.13 AFIHPSO optimization of the 2-dimensional *Schaffer's f6* function

f6	Fitness				Diversity mean	Period mean value
Inertia factor ω	Mean	Variance	Optimal value	Worst value		
0.9	−1	0	−1	−1	8.787×10^{-5}	610.733
0.5	−0.992	0.0155	−1	−0.915	2.902×10^{-3}	247.6
0.1	−0.997	0.0260	−1	−0.915	3.227×10^{-4}	396.6
−0.1	−1	0	−1	−1	4.788×10^{-3}	137.033
−0.5	−0.997	0.0105	−1	−0.915	9.584×10^{-4}	239.533
−0.9	−1	0	−1	−1	1.540×10^{-3}	2495.6

Table 5.14 Standard PSO optimization of the 30-dimensional *Schaffer's f6* function

f6	Fitness				Diversity mean	Period mean value
Inertia factor ω	Mean	Variance	Optimal value	Worst value		
0.9	−0.932	0.0318	−0.993	−0.899	5.459×10^{8}	3000
0.5	−1	2.873×10^{-9}	−1	−1	1.391×10^{-3}	83.115
0.1	−0.999	0.00103	−1	−0.995	6.204×10^{-4}	267.65
−0.1	−1	2.990×10^{-9}	−1	−1	9.372×10^{6}	172.55
−0.5	NaN	NaN	NaN	NaN	NaN	3000
−0.9	NaN	NaN	NaN	NaN	NaN	3000

5.3.3 Global Convergence of Harmonious Particle Swarm Optimization (HPSO) Algorithms

Bergh [16] demonstrated that the standard PSO cannot guarantee global or local convergence. He proposed the guaranteed convergence particle swarm optimization (GCPSO) and proved that GCPSO can ensure convergence to a *lbest*. Solis and Wets[15] provided a unified framework for the conditions and theorems of global convergence in random search algorithms. This section follows the approach in Gao et al. [15] to prove the global convergence of HPSO introduced in this chapter. First, the following assumptions and definitions are presented.

Fig. 5.13 Results of optimization iteration with AFIHPSO

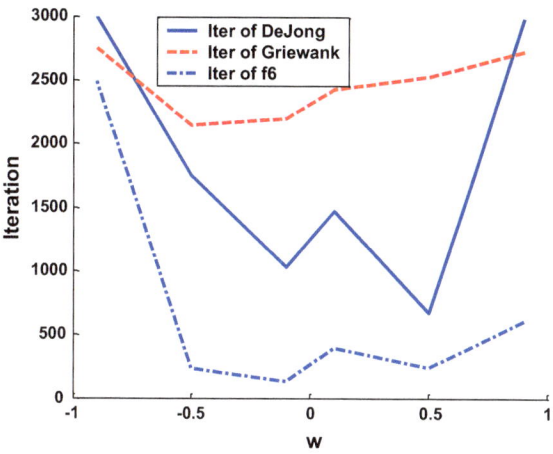

Fig. 5.14 Results of optimization iteration with standard PSO

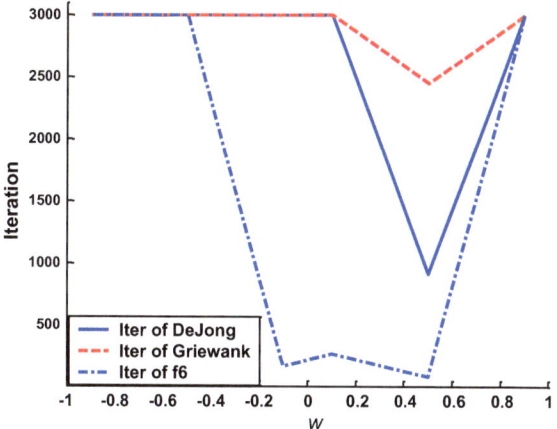

Assumptions:

1 The fitness function $f : S^n \to R_d$ is a single-valued function, and the search space S^n is finite;

2 $\alpha = \inf\{t : \upsilon[x \in S^n | f(x) < \varepsilon] > 0\}$, i.e., there exists a non-zero measure around the essential infimum α, where υ is the non-negative *Lebesgue* measure defined on the *Borel* set B in R^n and satisfies $\upsilon(S) > 0$.

Condition 5.1 In the sampling probability space (R^n, B, μ_k), generate $\xi^k \in S^n$, and $x^{k+1} = D(x^k, \xi^k)$. If $f(D(x, \xi)) \leq f(x)$, then $f(D(x, \xi)) \leq f(\xi)$. Here, $D : S^n \times R^n \to S^n$ is a mapping, S^n is a subset of R^n, B is the σ-algebra of R^n, and μ_k is a probability measure on B.

Condition 5.2 For any *Borel* subset A of S^n, if $\upsilon(A) > 0$, then $\prod_{k=0}^{\infty} [1 - \mu_k(A)] = 0$.

Theorem 5.2 *(Theorem of global search): Suppose that f is a measurable function, S^n is a measurable subset of R^n, and $\{x^k\}_{k=0}^{\infty}$ is a sequence generated by the algorithm. If both Condition 5.1 and Condition 5.2 are satisfied, then the following conclusion can be derived:*

$$\lim_{k \to \infty} P\left[x^k \in R_{\varepsilon,M}\right] = 1$$

where $P\left[x^k \in R_{\varepsilon,M}\right]$ represents the probability that the solution x^k produced by the algorithm at step k lies within $R_{\varepsilon,M}$.

Definition 5.1 *The optimal region $R_{\varepsilon,M}$ is defined as follows:*

$$R_{\varepsilon,M} = \begin{cases} \{x \in S | f(x) < \alpha + \varepsilon\} & \text{if } \alpha \text{ is finite} \\ \{x \in S | f(x) < M\} & \text{if } \alpha = -\infty \end{cases}$$

where $\varepsilon > 0$ and $M < 0$.

Based on above definitions and assumptions, the following lemma is derived.

Lemma 5.2 *AFHPSO is globally convergent.*

Proof We define the PSO in the form of the following D function, which satisfies Condition 5.1.

$$D\left(p_{gd}^k, x_{id}^k\right) = \begin{cases} p_{gd}^k & f\left(p_{gd}^k\right) \le f\left(x_{id}^k\right) \\ x_{id}^k & f\left(p_{gd}^k\right) > f\left(x_{id}^k\right) \end{cases}$$

The support set of particle i at the time step k, denoted as M_i^k is defined as:

$$M_i^{k+1} = x_{id}^k + \eta \cdot w \cdot v_{id}^k + \eta \cdot \varphi^{k+1} \cdot \left(p - x_{id}^k\right)$$

Then, M_i^{k+1} is a hypersphere centered at x_{id}^k with a radius of $\rho_i^{k+1} = \eta \cdot w \cdot v_{id}^k + \eta \cdot \varphi^k \cdot \left(p - x_{id}^k\right)$.

For standard PSO, except in some very special cases, we generally have $\upsilon\left(\bigcup_{i=1}^s M_i^{k+1} \cap S\right) < \upsilon(S)$. Thus, $\exists A \subset S$ such that $\sum_{i=0}^s \mu_i^k(A) = 0$. Therefore, standard PSO does not satisfy Condition 5.2, which implies that it cannot guarantee global convergence.

In AFHPSO, the search process undergoes a "contraction–expansion-contraction" cycle. The range of φ^{k+1} is not restricted, and the only factor influencing stability is $\Delta \eta \cdot \varphi^{k+1}$. Thus, theoretically, the range of ρ_i^{k+1} is not limited.

Based on the results from Lemma 5.1, the parameters can be adjusted if the following conditions are satisfied:

$$\eta \cdot \varphi^{k+1} \ge \frac{\left\| S_{\max}^n, S_{\min}^n \right\|_2 - \eta \cdot w \cdot v_{id}^k}{\left(p - x_{id}^k\right)}$$

Or

$$\eta \cdot w \geq \frac{\left\| S_{\max}^n, S_{\min}^n \right\|_2 - \eta \cdot \varphi^{k+1} \cdot \left(p - x_{id}^k \right)}{v_{id}^k}$$

Then we have $\rho_i^k \geq \left\| S_{\max}^n, S_{\min}^n \right\|_2$, where $S_{\max}^n$ and $S_{\min}^n$ are the upper and lower bounds of the search space S^n, respectively.

For M_i^k, since $DV_{swarm} \geq DV_{\max}$, the value of $\eta \cdot \varphi^{k+1}$ can satisfy Eq. (4. 4). Therefore, we have $S^n \subset M_i^k$, which further leads to $S^n \subset \bigcup_{i=1}^s M_i^k$ and $\upsilon \left(\bigcup_{i=1}^s M_i^k \cap S \right) = \upsilon(S)$. Consequently, $\forall A \subset S^n$, there exists $\prod_{k=0}^{\infty} [1 - \mu_k(A)] = 0$, thus, Condition 5.2 is satisfied.

With both Conditions 5.1 and 5.2 met, and according to Theorem 5.2, HPSO is globally convergent.

5.4 Summary

This chapter first explores the relationship between parameters and eigenvalue distribution when treating PSO as a time-invariant system. Next, it analyzes PSO as a time-varying system, examining the sufficient conditions for model stability in the absence of the *Lipschitz* condition. The results of Lemma 5.1 and Corollary 5.1 reduce one of the constraints of standard PSO and provide a method for selecting the acceleration factor φ^{k+1}. These results also expand the range of the inertia weight ω to $(-1, 1)$. The physical interpretation of selecting ω within $(-1, 0)$ can be understood as the particles consistently searching in the opposite direction. Based on the findings of Lemma 5.1 and Corollary 5.1, IFHPSO and AFHPSO were designed. Simulation experiments were then conducted to verify these results. Finally, the global convergence proof of the HPSO algorithms is provided.

References

1. Trelea IC (2003) The particle swarm optimization algorithm: convergence analysis and parameter selection [J]. Inf Process Lett 85(6):317–325
2. Clerc M, Kennedy J (2002) The particle swarm - explosion, stability, and convergence in a multidimensional complex space [J]. IEEE Trans Evol Comput 6(1):58–73
3. Zeng JC, Cui ZH (2006) A new unified model of particle swarm optimization and its theoretical analysis [J]. J Comput Res Dev 1:96–100
4. Cui ZH, Zeng JC (2006) Analysis and improvement about particle swarm optimization based on linear control theory [J]. J Chinese Comput Syst 5:849–853
5. Kadirkamanathan V, Selvarajah K, Fleming PJ (2006) Stability analysis of the particle dynamics in particle swarm optimizer [J]. IEEE Trans Evol Comput 10(3):245–255

6. Fernández-Martínez JL, García-Gonzalo E, Saraswathi S, Jernigan R, Kloczkowski A (2011) Particle swarm optimization: a powerful family of stochastic optimizers. Anal Des Appl Inverse Model [J] 1–8

7. Fernandez-Martinez JL, Garcia-Gonzalo E, Saraswathi S, et al (2011) Particle swarm optimization: a powerful family of stochastic optimizers. Analysis, design and application to inverse modelling. In: Proceedings of the 2nd International conference on swarm intelligence, ICSI 2011, June 12, 2011–June 15, 2011. Springer Verlag, Chongqing, China

8. Pan F, Chen J, Gan GM, et al (2006) Model analysis of particle swarm optimizer [J]. Acta Automat Sin 3:368–377

9. Clerc M (2006) Stagnation analysis in particle swarm optimization or what happens when nothing happens

10. Ming J, Yupin L, Shiyuan Y (2007) Stagnation analysis in particle swarm optimization. In: Proceedings of the 2007 IEEE swarm intelligence symposium. IEEE, Piscataway, NJ, USA

11. Poli R, Broomhead D (2007) Exact analysis of the sampling distribution for the canonical particle swarm optimiser and its convergence during stagnation. In: Proceedings of the 9th annual genetic and evolutionary computation conference, GECCO 2007, July 7, 2007–July 11, 2007. Association for Computing Machinery, London, United kingdom

12. Pan F, Chen J, Xin B, et al (2009) Several characteristics analysis of particle swarm optimizer [J]. Acta Automat Sin 7:1010–1016

13. Zhan ZH, Li JJ, Cao JN (2013) Multiple populations for multiple objectives: a coevolutionary technique for solving multiobjective optimization problems[J]. IEEE Trans Cybernet 43(2):445–463

14. Wan FC, Wang DW, Li YP (2004) Particls swarm optimization of correlative product combinatorial introduction model [J]. Control Decis 5:520–524

15. Gao S, Han B, Wu XJ, et al (2004) Solving traveling salesman problem by hybrid particle swarm optimization algorithm [J]. Control Decis 11:1286–1289

16. van den Bergh F (2002) An analysis of particle swarm optimizers [D]. University of Pretoria, South Africa

Chapter 6
Markov Chain Analysis of the Standard PSO

One approach to studying the convergence of PSO is through the general framework of stochastic search algorithms. From the perspective of stochastic processes, Jin [1] transformed the dynamic model of PSO with random factors into a linear time-invariant system for probabilistic analysis, providing a sufficient condition for the system's convergence in probability. Inspired by Markov chain theory, Yuan [2] introduced the "forgetting characteristic" to particles in PSO, establishing a mathematical model for the algorithm's Markov chain process. Li [3] demonstrated that the chain of particle states in PSO is a Markov chain and that the sequence of the swarm state is a finite homogeneous Markov chain. Additionally, Li analyzed the particle trajectories from the standpoint of difference equations. Ren [4] focused on the Markov chain formed by particle states, proving the reducibility and non-homogeneity of the state space. He confirmed that the state space is transient and, ultimately, demonstrated the conditions under which a stationary process does not exist in the Markov chain, thereby concluding that global convergence is not guaranteed. Cai [5] developed an absorbing Markov process for PSO and proposed the reachable state set as a critical indicator for convergence analysis, also suggesting methods for enhancing the global convergence of PSO. With a finite element grid, Poli [6] constructed a discrete Markov chain model for BBPS and showed that this model can approximate continuous problems to any desired level of accuracy. The transition matrix of the iterative chain provides precise information about particle behavior in each generation of BBPS, including the probability of finding the gbest or encountering incorrect solutions. *Poli* also explained the characteristics of applying Cauchy, Gaussian, and other probability distributions to the acceleration factors in standard PSO.

Based on the difference model of the standard PSO algorithm, this chapter analyzes the sequence properties of particles and swarm state, examining issues such as premature convergence, divergence, and the conditions under which the standard PSO converges to *gbest* with a certain probability. The overall structure is depicted in Fig. 6.1. First, with the difference model of the PSO algorithm, the

© Beijing Institute of Technology Press 2025
F. Pan et al., *Particle Swarm Optimizer and Multi-Objective Optimization*,
https://doi.org/10.1007/978-981-95-3381-7_6

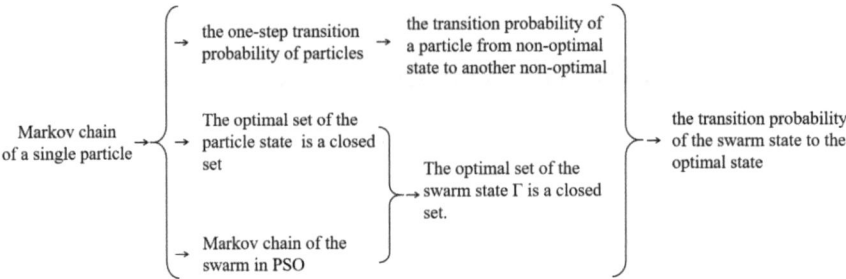

Fig. 6.1 The overall structure of the analysis and proof in this section

sequence states for the particles and the swarm are defined, followed by an analysis of their Markov properties. The closedness of the optimal state set of particles and the swarm is proven, and the one-step transition probability of particles is calculated. Furthermore, based on the total probability formula and properties of Markov chain, the transition probability of the swarm state to the optimal state set is derived. With this transition probability, the inertia weight ω and acceleration factor c of the PSO algorithm are discussed and interpreted, and issues such as premature convergence and divergence are investigated. Finally, the analysis concludes that standard PSO converges to *gbest* with a certain probability.

6.1 Markov Chain Analysis of a Single Particle in Standard PSO

At any given time step k, the update equations of velocity and position for particle i are defined as:

$$\begin{cases} v_i^{(k+1)} = \omega \cdot v_i^{(k)} + c_1 \cdot r_{i1}^{(k)} \cdot (p_i^{(k)} - x_i^{(k)}) + c_2 \cdot r_{i2}^{(k)} \cdot (p_l^{(k)} - x_i^{(k)}) \\ x_i^{(k+1)} = x_i^{(k)} + v_i^{(k+1)} \end{cases} \tag{6.1}$$

where $i = 1, 2, \ldots, m$, m represents the total number of particles; ω is the inertia weight, $x_i^{(k)}$ represents the current position vector, $v_i^{(k)}$ represents the velocity vector, $p_i^{(k)}$ represents the *pbest* position, and $p_l^{(k)}$ represents the *gbest* position within the swarm. The parameters c_1 and c_2 are acceleration factors, typically set as $c_1 = c_2 = c$, while $r_{i1}^{(k)}$ and $r_{i2}^{(k)}$ are random numbers uniformly distributed between $[0, 1]$. Based on Eq. (6.1), the difference model of PSO can be derived as follows:

$$\begin{aligned} x_i^{(k+1)} &= x_i^{(k)} + v_i^{(k+1)} \\ &= x_i^{(k)} + \omega \cdot (x_i^{(k)} - x_i^{(k-1)}) + \varphi_{i1}^k \cdot (p_i^{(k)} - x_i^{(k)}) + \varphi_{i2}^k \cdot (p_l^{(k)} - x_i^{(k)}) \end{aligned} \tag{6.2}$$

where $\varphi_{i1}^{(k)} = c_1 \cdot r_{i1}^{(k)}$, $\varphi_{i2}^{(k)} = c_2 \cdot r_{i2}^{(k)}$.

Definition 6.1 Let $\{y^{(k)}; k \geq 0\}$ be a sequence of discrete random variables, with a set of possible discrete values $S = \{j\}$, called the state space. If, for any $k \geq 1$, $i^{(l)} \in S (l \leq k + 1)$ satisfies:

$$p(y^{(k+1)} = i^{(k+1)}|y^{(k)} = i^{(k)}, \cdots, y^{(0)} = i^{(0)}) = p(y^{(k+1)} = i^{(k+1)}|y^{(k)} = i^{(k)}) \quad (6.3)$$

Then $\{y^{(k)}; k \geq 0\}$ is a Markov chain.

Definition 6.2 According to Eq. (6.2), the state of particle i in the swarm under PSO at time step k is defined as $\xi_i^{(k)}$, i.e. $\xi_i^{(k)} = (x_i^{(k)}, x_i^{(k-1)}, p_i^{(k)}, p_l^{(k)})$.

Lemma 6.1 The sequence $\{\xi_i^{(k)}, k \geq 1\}$ formed by the state $\xi_i^{(k)}$ of particle i in the PSO at time step k is a Markov chain.

Proof By the definition in Eq. (6.2), $\{\xi_i^{(k)}, k \geq 1\}$ is a sequence of discrete random variables, and the value of $\xi_i^{(k+1)} = (x_i^{(k+1)}, x_i^{(k)}, p_i^{(k+1)}, p_l^{(k+1)})$ depends solely on the state $\xi_i^{(k)} = (x_i^{(k)}, x_i^{(k-1)}, p_i^{(k)}, p_l^{(k)})$ at the previous time step. Let the state space $\{\xi_i^{(k)}, k \geq 1\}$ be S. For any $k \geq 1$ and $\xi_i^{(l)} \in S (l \leq k + 1)$, we have:

$$p(\xi^{(k+1)}|\xi^{(k)}, \cdots, \xi^{(1)}) = p(\xi^{(k+1)}|\xi^{(k)}) \quad (6.4)$$

Therefore, the sequence $\{\xi_i^{(k)}, k \geq 1\}$ formed by the state $\xi_i^{(k)} = (x_i^{(k)}, x_i^{(k-1)}, p_i^{(k)}, p_l^{(k)})$ of particle i at time step k in PSO is a Markov chain. $\square$

Lemma 6.2 When the acceleration factors $\varphi_{i1}^{(k)}, \varphi_{i2}^{(k)} \sim U(0, c)$ in PSO, the one-step transition probability for particle i moving from $x_i^{(k)}$ to a spherical region centered at $x_i^{(k+1)}$ with a radius of ε is:

$$p(\xi_i^{(k+1)}|\xi_i^{(k)}) = \frac{\varepsilon}{\omega \cdot ||x_i^{(k)} - x_i^{(k-1)}||} \cdot \frac{\varepsilon}{c \cdot ||p_i^{(k)} - x_i^{(k)}||} \cdot \frac{\varepsilon}{c \cdot ||p_l^{(k)} - x_i^{(k)}||} \quad (6.5)$$

where, If $|f(p_i^{(k)}) - f(x_i^{(k)})| \leq \delta$, then $\frac{\varepsilon}{c \cdot ||p_i^{(k)} - x_i^{(k)}||} = 1$, and if $|f(p_l^{(k)}) - f(x_i^{(k)})| \leq \delta$, then $\frac{\varepsilon}{c \cdot ||p_l^{(k)} - x_i^{(k)}||} = 1$ (δ is a value converges to 0).

Proof Consider the single-particle model:

$$x_i^{(k+1)} = x_i^{(k)} + \omega \cdot (x_i^{(k)} - x_i^{(k-1)}) + \varphi_{i1}^k \cdot (p_i^{(k)} - x_i^{(k)}) + \varphi_{i2}^k \cdot (p_l^{(k)} - x_i^{(k)}) \quad (6.6)$$

Since $\varphi_{i1}^{(k)}$ and $\varphi_{i2}^{(k)}$ are random numbers, the value of $x_t^{(k+1)}$ is determined by $\omega, \varphi_{i1}^{(k)}$ and $\varphi_{i2}^{(k)}$. Let's examine the following three scenarios:

① When $|f(p_i^{(k)}) - f(x_i^{(k)})| \leq \delta$ and $|f(p_l^{(k)}) - f(x_i^{(k)})| \leq \delta$, as shown in Fig. 6.2a, we have:

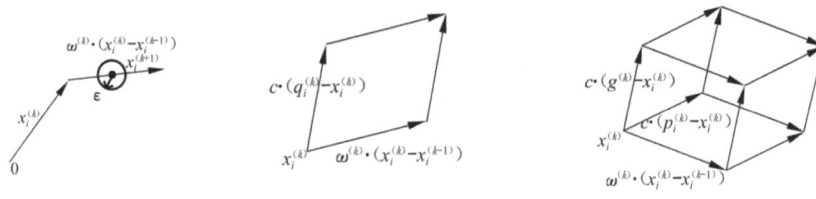

(a) The first scenario   (b) The second scenario   (c) The third scenario

Fig. 6.2 One-step movement range of a single particle

$$x_i^{(k+1)} = x_i^{(k)} + \omega \cdot (x_i^{(k)} - x_i^{(k-1)}) \tag{6.7}$$

In this scenario, the endpoint $x_i^{(k+1)}$ lies on the vector $\omega \cdot (x_i^{(k)} - x_i^{(k-1)})$. Therefore, $x_i^{(k+1)}$ can be uniquely determined by ω. When ω is an unknown constant, we have:

$$p\left(x_i^{(k+1)}|\xi_i^{(k)}\right) = \frac{\int_{x_i^{(k+1)}-\frac{1}{2}\varepsilon}^{x_i^{(k+1)}+\frac{1}{2}\varepsilon} dy}{\int_{x_i^{(k)}}^{x_i^{(k)}+\omega\cdot\left(x_i^{(k)}-x_i^{(k-1)}\right)} dy} = \frac{\varepsilon}{\omega \cdot \left\| x_i^{(k)} - x_i^{(k-1)} \right\|} \tag{6.8}$$

② When $|f(p_i^{(k)}) - f(x_i^{(k)})| \leq \delta$ or $|f(p_l^{(k)}) - f(x_i^{(k)})| \leq \delta$, Let $q_i^{(k)} \in \left\{ p_i^{(k)}, p_l^{(k)} \right\}$, as shown in Fig. 6.2b, we have:

$$x_i^{(k+1)} = x_i^{(k)} + \omega \cdot (x_i^{(k)} - x_i^{(k-1)}) + \varphi_i^k \cdot (q_i^{(k)} - x_i^{(k)}) \tag{6.9}$$

where $\varphi_i^{(k)}$ is a random number. Thus, $x_i^{(k+1)}$ is located within the region defined by both $\omega \cdot (x_i^{(k)} - x_i^{(k-1)})$ and $\varphi_i^k \cdot (q_i^{(k)} - x_i^{(k)})$. Consequently, the value of $x_i^{(k+1)}$ is determined by both ω and $\varphi_i^{(k)}$, thus, we have:

$$p\left(x_i^{(k+1)}|\xi_i^{(k)}\right) = \frac{\int_{x_i^{(k+1)}-\frac{1}{2}\varepsilon}^{x_i^{(k+1)}+\frac{1}{2}\varepsilon} dy}{\int_{x_i^{(k)}}^{x_i^{(k)}+\omega\cdot\left(x_i^{(k)}-x_i^{(k-1)}\right)} dy} \cdot \frac{\int_{\varphi_i^{(k)}-\frac{1}{2}\varepsilon}^{\varphi_i^{(k)}+\frac{1}{2}\varepsilon} dy}{\int_{x_i^{(k)}}^{x_i^{(k)}+c\cdot(q_i^{(k)}-x_i^{(k)})} dy}$$

$$= \frac{\varepsilon}{\omega \cdot \left\| x_i^{(k)} - x_i^{(k-1)} \right\|} \cdot \frac{\varepsilon}{c \cdot \left\| q_i^{(k)} - x_i^{(k)} \right\|} \tag{6.10}$$

③ When $|f(p_i^{(k)}) - f(x_i^{(k)})| > \delta$ and $|f(p_l^{(k)}) - f(x_i^{(k)})| > \delta$, as shown in Fig. 6.2c, we have:

$$x_i^{(k+1)} = x_i^{(k)} + \omega \cdot (x_i^{(k)} - x_i^{(k-1)}) + \varphi_{i1}^k \cdot (p_i^{(k)} - x_i^{(k)}) + \varphi_{i2}^k \cdot (p_l^{(k)} - x_i^{(k)}) \quad (6.11)$$

When $x_i^{(k+1)} \in R$, $x_i^{(k+1)}$ can only be determined when ω, $\varphi_{i1}^{(k)}$ and $\varphi_{i2}^{(k)}$ are all specified. Therefore:

$$p\left(x_i^{(k+1)} | \xi_i^{(k)}\right) = \frac{\int_{x_i^{(k+1)} - \frac{1}{2}\varepsilon}^{x_i^{(k+1)} + \frac{1}{2}\varepsilon} dy}{\int_{x_i^{(k)}}^{x_i^{(k)} + \omega \cdot \left(x_i^{(k)} - x_i^{(k-1)}\right)} dy} \cdot \frac{\int_{\varphi_{i1}^{(k)} - \frac{1}{2}\varepsilon}^{\varphi_{i1}^{(k)} + \frac{1}{2}\varepsilon} dy}{\int_{x_i^{(k)}}^{x_i^{(k)} + c \cdot \left(p_i^{(k)} - x_i^{(k)}\right)} dy} \cdot \frac{\int_{\varphi_{i2}^{(k)} - \frac{1}{2}\varepsilon}^{\varphi_{i2}^{(k)} + \frac{1}{2}\varepsilon} dy}{\int_{x_i^{(k)}}^{x_i^{(k)} + c \cdot \left(p_l^{(k)} - x_i^{(k)}\right)} dy}$$

$$= \frac{\varepsilon}{\omega \cdot \left\| x_i^{(k)} - x_i^{(k-1)} \right\|} \cdot \frac{\varepsilon}{c \cdot \left\| p_i^{(k)} - x_i^{(k)} \right\|} \cdot \frac{\varepsilon}{c \cdot \left\| g^{(k)} - x_i^{(k)} \right\|} \quad (6.12)$$

Combing ①, ②, and ③, we have:

$$p\left(x_i^{(k+1)} | \xi_i^{(k)}\right) = \frac{\varepsilon}{\omega \cdot \left\| x_i^{(k)} - x_i^{(k-1)} \right\|} \cdot \frac{\varepsilon}{c \cdot \left\| p_i^{(k)} - x_i^{(k)} \right\|} \cdot \frac{\varepsilon}{c \cdot \left\| p_l^{(k)} - x_i^{(k)} \right\|} \quad (6.13)$$

$$p\left(x_i^{(k+1)} | \xi_i^{(k)}\right) = \frac{\varepsilon}{\omega \cdot \left\| x_i^{(k)} - x_i^{(k-1)} \right\|} \cdot \frac{\varepsilon}{c \cdot \left\| p_i^{(k)} - x_i^{(k)} \right\|} \cdot \frac{\varepsilon}{c \cdot \left\| p_l^{(k)} - x_i^{(k)} \right\|} \quad (6.14)$$

Now we have $\lim\limits_{p_i^{(k)} \to x_i^{(k)}} \frac{\varepsilon}{c \cdot \left\| p_i^{(k)} - x_i^{(k)} \right\|} = 1$, $\lim\limits_{g^{(k)} \to x_i^{(k)}} \frac{\varepsilon}{c \cdot \left\| p_i^{(k)} - x_i^{(k)} \right\|} = 1$.

Since $p_i^{(k+1)}$ and $p_l^{(k+1)}$ are determined by $x_i^{(k+1)}$, the probability of particle i transitioning from state $\xi^{(k)}$ to $\xi^{(k+1)}$ is:

$$p\left(\xi_i^{(k+1)} | \xi_i^{(k)}\right) = \frac{\varepsilon}{\omega \cdot \left\| x_i^{(k)} - x_i^{(k-1)} \right\|} \cdot \frac{\varepsilon}{c \cdot \left\| p_i^{(k)} - x_i^{(k)} \right\|} \cdot \frac{\varepsilon}{c \cdot \left\| p_l^{(k)} - x_i^{(k)} \right\|} \quad (6.15)$$

where ε represents the radius of the region in which $x_i^{(k+1)}$ is located. When $|f(p_i^{(k)}) - f(x_i^{(k)})| \leq \delta$, $\frac{\varepsilon}{c \cdot \left\| p_i^{(k)} - x_i^{(k)} \right\|} = 1$, and when $|f(p_l^{(k)}) - f(x_i^{(k)})| \leq \delta$, $\frac{\varepsilon}{c \cdot \left\| p_l^{(k)} - x_i^{(k)} \right\|} = 1$ (δ is a value converges to 0). $\qquad \square$

Definition 6.3 Optimal set of the particle state. Let the gbest of the optimization problem be g^*. The optimal set of the particle state is defined as: $M = \{\xi^* = (x^{(k)}, x^{(k-1)}, g^*, g^*), k \geq 1\}$, where $\xi^* \in S$ represents the optimal state. If $M = S$, every solution within the feasible region is an optimal solution, which means the optimization problem is resolved. Therefore, we consider the case where $M \subset S$.

Definition 6.4 A subset H of the state space S is a closed set if, for any state $i \in H$: $\sum_{j \in H} p_{ij}$. That is, the state i in H cannot reach any state outside of H: $p(i + 1 \in H | i \in H) = 1$.

Lemma 6.3 The optimal set of the particle state M is a closed set.

Proof According to the selection strategy of the standard PSO, when the particle state $\xi_i^{(k)} = (x_i^{(k)}, x_i^{(k-1)}, p_i^{(k)}, g^{(k)})$ is an optimal state, i.e., $\xi_i^{(k)} = \xi^* = (x^{(k)}, x^{(k-1)}, g^*, g^*)$, the state at the next time step $\xi_i^{(k+1)} = (x^{(k+1)}, x^{(k)}, g^*, g^*)$ will also be an optimal state. Therefore, if $\xi_i^{(k)} \in M$, $k \in [0, T)$, then $\forall l \in [k + 1, T]$, $\xi_i^{(l)} \in M$. Thus:

$$\begin{cases} P(\xi_i^{(k+1)} \in M \mid \xi_i^{(k)} \in M) = 1 \\ P(\xi_i^{(k+1)} \notin M \mid \xi_i^{(k)} \in M) = 0 \end{cases} \tag{6.16}$$

Hence, the optimal set of the particle state M is a closed set.

Lemma 6.4 For the state $\xi_i^{(k)} \notin M$ at time step k, the probability that its next state $\xi_i^{(k+1)} \notin M$ is given by:

$$p\left(\xi_i^{(k+1)} \notin M \mid \xi_i^{(k)} \notin M\right)$$
$$= 1 - \frac{\varepsilon_k}{\omega \cdot \left\| x_i^{(k)} - x_i^{(k-1)} \right\|} \cdot \frac{\varepsilon_k}{c \cdot \left\| p_i^{(k)} - x_i^{(k)} \right\|} \cdot \frac{\varepsilon_k}{c \cdot \left\| p_l^{(k)} - x_i^{(k)} \right\|} \tag{6.17}$$

where, if the gbest $g^* \in R$, then $\varepsilon_k = \varepsilon$; if $g^* \notin R$, then $\varepsilon_k = 0$.

Proof Analyzing the motion pattern of particles in the standard PSO reveals that the particle state $\xi_i^{(k)} \notin M$ is known at time step k. Due to the uncertainty of the acceleration factors $\varphi_{i1}^{(k)}$ and $\varphi_{i2}^{(k)}$, the position of the particle at time $k + 1$ lies within a feasible region R. Hence, the location of the *gbest* g^* is unknown; it may lie either inside or outside the region R. According to Lemma 6.2, at time step k, $\forall \xi_i^{(k+1)} \in R(k \geq 1)$, we have:

$$p\left(\xi_i^{(k+1)} \mid \xi_i^{(k)}\right) = \frac{\varepsilon}{\omega \cdot \left\| x_i^{(k)} - x_i^{(k-1)} \right\|} \cdot \frac{\varepsilon}{c \cdot \left\| p_i^{(k)} - x_i^{(k)} \right\|} \cdot \frac{\varepsilon}{c \cdot \left\| p_l^{(k)} - x_i^{(k)} \right\|} \tag{6.18}$$

And:

$$p\left(\xi_i^{(k+1)} \in M \mid \xi_i^{(k)} \notin M\right)$$
$$= \begin{cases} 0 & g^* \notin R \\ \dfrac{\varepsilon}{\omega \cdot \left\| x_i^{(k)} - x_i^{(k-1)} \right\|} \cdot \dfrac{\varepsilon}{c \cdot \left\| p_i^{(k)} - x_i^{(k)} \right\|} \cdot \dfrac{\varepsilon}{c \cdot \left\| p_l^{(k)} - x_i^{(k)} \right\|} & g^* \in R \end{cases} \tag{6.19}$$

Since:

$$P(\xi_i^{(k+1)} \notin M \mid \xi_i^{(k)} \notin M) = 1 - P(\xi_i^{(k+1)} \in M \mid \xi_i^{(k)} \notin M) \tag{6.20}$$

Thus we have:

$$p\left(\xi_i^{(k+1)} \notin M \mid \xi_i^{(k)} \notin M\right)$$

$$= \begin{cases} 1 & g^* \notin R \\ 1 - \dfrac{\varepsilon}{\omega \cdot \left\| x_i^{(k)} - x_i^{(k-1)} \right\|} \cdot \dfrac{\varepsilon}{c \cdot \left\| p_i^{(k)} - x_i^{(k)} \right\|} \cdot \dfrac{\varepsilon}{c \cdot \left\| p_l^{(k)} - x_i^{(k)} \right\|} & g^* \in R \end{cases} \qquad (6.21)$$

Combining Eq. (6.21), we get:

$$p\left(\xi_i^{(k+1)} \notin M \mid \xi_i^{(k)} \notin M\right)$$

$$= 1 - \dfrac{\varepsilon_k}{\omega \cdot \left\| x_i^{(k)} - x_i^{(k-1)} \right\|} \cdot \dfrac{\varepsilon_k}{c \cdot \left\| p_i^{(k)} - x_i^{(k)} \right\|} \cdot \dfrac{\varepsilon_k}{c \cdot \left\| p_l^{(k)} - x_i^{(k)} \right\|} \qquad (6.22)$$

where, if $g^* \in R$, $\varepsilon_k = \varepsilon$; if $g^* \notin R$, $\varepsilon_k = 0$. $\qquad\square$

6.2 Markov Chain Analysis of the Swarm in PSO

Definition 6.5 The state of the swarm in PSO is defined as: $\varsigma^{(k)} = \left(\xi_1^{(k)}, \xi_2^{(k)}, \cdots, \xi_m^{(k)}\right)$, where m represents the number of particles in the swarm, and k represents the current number of iterations.

Lemma 6.5 The state sequence of a PSO swarm that is composed of m particles, $\{\varsigma^{(k)} = (\xi_1^{(k)}, \xi_2^{(k)}, \cdots, \xi_m^{(k)}), k \geq 1\}$, forms a Markov chain.

Proof Let the state space of $\{\varsigma^{(k)}, k \geq 1\}$ be Λ, For any $k \geq 1$ and $\varsigma^{(l)} \in \Lambda (l \leq k+1)$, we have:

$$p(\varsigma^{(k+1)} \mid \varsigma^{(k)}, \cdots, \varsigma^{(k)}) = p(\xi_1^{(k+1)}, \cdots, \xi_m^{(k+1)} \mid \xi_1^{(k)}, \cdots, \xi_m^{(k)}, \cdots, \xi_1^{(1)}, \cdots, \xi_m^{(1)}) \qquad (6.23)$$

From Lemma 6.1, we know that $p(\xi^{(k+1)} \mid \xi^{(k)}, \cdots, \xi^{(k)}) = p(\xi^{(k+1)} \mid \xi^{(k)})$, hence we have:

$$p(\varsigma^{(k+1)} \mid \varsigma^{(k)}, \cdots, \varsigma^{(k)}) = p(\xi_1^{(k+1)}, \cdots, \xi_m^{(k+1)} \mid \xi_1^{(k)}, \cdots, \xi_m^{(k)}) = p(\varsigma^{(k+1)} \mid \varsigma^{(k)}) \qquad (6.24)$$

Therefore, the state sequence $\{\varsigma^{(k)} = (\xi_1^{(k)}, \xi_2^{(k)}, \cdots, \xi_m^{(k)}), k \geq 1\}$ with m particles constitutes a Markov chain. $\qquad\square$

Definition 6.6 The optimal set of the swarm state Γ. The swarm state is $\varsigma^{(k)} = (\xi_1^{(k)}, \xi_2^{(k)}, \cdots, \xi_m^{(k)})$, the gbest state is defined as ς_k^*, for $\varsigma_k^* =$

$(\xi_1^{(k)}, \cdots, \xi_i^{(k)}, \cdots, \xi_m^{(k)})$, $\exists i \in [1, m]$, such that $\xi_i^{(k)} \in M$ satisfies the optimal set of the swarm state

$$\Gamma = \{\varsigma_k^*, k \geq 1 | \varsigma_k^* = (\xi_1^{(k)}, \cdots, \xi_i^{(k)}, \cdots, \xi_m^{(k)}), \exists i \in [1, m], \xi_i^{(k)} \in M\}$$

If the optimal set of the state $\Gamma = \Lambda$ (the state space of the swarm), this indicates that all the states of the PSO swarm are optimal, implying that the algorithm has reached convergence. Thus, we consider the case where $\Gamma \subset \Lambda$.

Lemma 6.6 The optimal set of the swarm state Γ is a closed set.

Proof From Lemma 6.3, we know that the optimal state set M is a closed set. Thus, we have:

$$P(\xi_i^{(k+1)} \in M | \xi_i^{(k)} \in M) = 1 \tag{6.25}$$

And when $\Gamma = \{\varsigma_k^*, k \geq 1 | \varsigma_k^* = (\xi_1^{(k)}, \cdots, \xi_i^{(k)}, \cdots, \xi_m^{(k)}), \exists i \in [1, m], \xi_i^{(k)} \in M\}$, if at time step k, the swarm state $\varsigma^{(k)} = (\xi_1^{(k)}, \xi_2^{(k)}, \cdots, \xi_m^{(k)}) \in \Gamma$, that is, for the swarm state $\varsigma^{(k)}$, $\exists i \in [1, m]$, $\xi_i^{(k)} \in M$. Then at time step $k + 1$, it must also hold that $\xi_i^{(k+1)} \in M$. Consequently, the swarm state at time step $k + 1$, $\varsigma^{(k+1)} = (\xi_1^{(k+1)}, \cdots, \xi_m^{(k+1)})$, will satisfy $\exists i \in [1, m]$, $\xi_i^{(k+1)} \in M$. Therefore, $\varsigma^{(k+1)} \subset \Gamma$, which implies:

$$\begin{cases} p(\varsigma^{(k+1)} \in \Gamma | \varsigma^{(k)} \in \Gamma) = 1 \\ p(\varsigma^{(k+1)} \notin \Gamma | \varsigma^{(k)} \in \Gamma) = 0 \end{cases} \tag{6.26}$$

Thus, the optimal state set Γ is a closed set. $\qquad \square$

Lemma 6.7 The probability that the swarm state $\varsigma^{(k+1)} \in \Gamma$ at time step $k + 1$ is:

$$P(\varsigma^{(k+1)} \in \Gamma)$$
$$= 1 - p(\varsigma^{(1)} \notin \Gamma)$$
$$\cdot \prod_{l=1}^{k} \prod_{i=1}^{m} \left(1 - \frac{\varepsilon_l}{\omega \cdot ||x_i^{(k)} - x_i^{(k-1)}||} \cdot \frac{\varepsilon_l}{c \cdot ||p_i^{(k)} - x_i^{(k)}||} \cdot \frac{\varepsilon_l}{c \cdot ||p_l^{(k)} - x_i^{(k)}||}\right) \tag{6.27}$$

Proof According to the law of total probability, the probability that the swarm state $\varsigma^{k+1} \notin \Gamma$ at time step $k + 1$ is:

$$p(\varsigma^{(k+1)} \notin \Gamma) = p(\varsigma^{(k+1)} \notin \Gamma | \varsigma^{(k)} \notin \Gamma) \cdot p(\varsigma^{(k)} \notin \Gamma)$$
$$+ p(\varsigma^{(k+1)} \notin \Gamma | \varsigma^{(k)} \in \Gamma) \cdot p(\varsigma^{(k)} \in \Gamma) \tag{6.28}$$

And since: $p(\varsigma^{(k+1)} \notin \Gamma | \varsigma^{(k)} \in \Gamma) = 0$.
Hence:

$$
\begin{aligned}
p(\varsigma^{(k+1)} \notin \Gamma) &= p(\varsigma^{(k+1)} \notin \Gamma | \varsigma^{(k)} \notin \Gamma) \cdot p(\varsigma^{(k)} \notin \Gamma) \\
&= p(\varsigma^{(k+1)} \notin \Gamma | \varsigma^{(k)} \notin \Gamma) \cdot p(\varsigma^{(k)} \notin \Gamma | \varsigma^{(k-1)} \notin \Gamma) \cdots \cdot p(\varsigma^{(1)} \notin \Gamma) \\
&= p(\varsigma^{(1)} \notin \Gamma) \cdot \prod_{l=1}^{k} p(\varsigma^{(l+1)} \notin \Gamma | \varsigma^{(l)} \notin \Gamma)
\end{aligned}
\tag{6.29}
$$

Given that the swarm state $\varsigma^{(k)} = (\xi_1^{(k)}, \xi_2^{(k)}, \cdots, \xi_m^{(k)}) \notin \Gamma$, this means that $\forall i \in [1, m], \overline{\exists}\, \xi_i^k \in M$, where $\overline{\exists}$ indicates non-existence.

Thus, we have:

$$
\begin{aligned}
&p\left(\varsigma^{(k+1)} \notin \Gamma | \varsigma^{(k)} \notin \Gamma\right) \\
&= \prod_{i=1}^{m} p\left(\xi_i^{(k+1)} \notin M | \xi_i^{(k)} \notin M\right) \\
&= \prod_{i=1}^{m}\left(1 - \frac{\varepsilon_k}{\omega \cdot \left\| x_i^{(k)} - x_i^{(k-1)} \right\|} \cdot \frac{\varepsilon_k}{c \cdot \left\| p_i^{(k)} - x_i^{(k)} \right\|} \cdot \frac{\varepsilon_k}{c \cdot \left\| p_l^{(k)} - x_i^{(k)} \right\|}\right)
\end{aligned}
\tag{6.30}
$$

Therefore:

$$
\begin{aligned}
&p(\varsigma^{(k+1)} \notin \Gamma) \\
&= p(\varsigma^{(1)} \notin \Gamma) \cdot \prod_{l=1}^{k} p(\varsigma^{(l+1)} \notin \Gamma | \varsigma^{(l)} \notin \Gamma) \\
&= p(\varsigma^{(1)} \notin \Gamma) \\
&\quad \cdot \prod_{l=1}^{k} \prod_{i=1}^{m}\left(1 - \frac{\varepsilon_l}{\omega \cdot \left\| x_i^{(k)} - x_i^{(k-1)} \right\|} \cdot \frac{\varepsilon_l}{c \cdot \left\| p_i^{(k)} - x_i^{(k)} \right\|} \cdot \frac{\varepsilon_l}{c \cdot \left\| p_l^{(k)} - x_i^{(k)} \right\|}\right)
\end{aligned}
$$

Hence, the probability that the swarm state $\varsigma^{(k+1)} \in \Gamma$ at time step $k + 1$ is:

$$
\begin{aligned}
&p(\varsigma^{(k+1)} \in \Gamma) \\
&= 1 - p(\varsigma^{(k+1)} \notin \Gamma) \\
&= 1 - p(\varsigma^{(1)} \notin \Gamma) \\
&\quad \cdot \prod_{l=1}^{k} \prod_{i=1}^{m}\left(1 - \frac{\varepsilon_l}{\omega \cdot \left\| x_i^{(k)} - x_i^{(k-1)} \right\|} \cdot \frac{\varepsilon_l}{c \cdot \left\| p_i^{(k)} - x_i^{(k)} \right\|} \cdot \frac{\varepsilon_l}{c \cdot \left\| p_l^{(k)} - x_i^{(k)} \right\|}\right)
\end{aligned}
\tag{6.31}
$$

6.3   Analysis of the Impact of PSO Parameters on Optimization Performance

6.3.1   Effect of Swarm Size

Consider Eq. (6.30). When $m \to \infty$(i.e., the number of particles in the swarm becomes infinite), since the particles are randomly distributed (independent of each other), and $1 - \frac{\varepsilon_k}{w \cdot \left\| x_i^{(k)} - x_i^{(k-1)} \right\|} \cdot \frac{\varepsilon_k}{c \cdot \left\| p_i^{(k)} - x_i^{(k)} \right\|} \cdot \frac{\varepsilon_k}{c \cdot \left\| p_l^{(k)} - x_i^{(k)} \right\|} < 1$, thus, we have:

$$\prod_{i=1}^{\infty} \left(1 - \frac{\varepsilon_k}{w \cdot \left\| x_i^{(k)} - x_i^{(k-1)} \right\|} \cdot \frac{\varepsilon_k}{c \cdot \left\| p_i^{(k)} - x_i^{(k)} \right\|} \cdot \frac{\varepsilon_k}{c \cdot \left\| p_l^{(k)} - x_i^{(k)} \right\|} \right) \to 0. \text{ This implies that } p(\varsigma^{(k+1)} \notin$$

$\Gamma | \varsigma^k \notin \Gamma) \to 0$, and thus $p(\varsigma^{(k+1)} \in \Gamma) \to 1$. This indicates that as the number of particles approaches infinity, the algorithm converges to gbest with a probability of 1, which is similar to performing an infinite number of random sampling in the solution space.

6.3.2   Effect of Inertia Weight Ω in PSO

Consider the equation $x_i^{(k+1)} = x_i^{(k)} + \omega \cdot (x_i^{(k)} - x_i^{(k-1)}) + \varphi_{i1}^{(k)} \cdot (p_i^{(k)} - x_i^{(k)}) + \varphi_{i2}^{(k)} \cdot (p_l^{(k)} - x_i^{(k)})$. The feasible region of particle i is determined by ω, $\varphi_{i1}^{(k)}$ and $\varphi_{i2}^{(k)}$. When $x_i^{(k)}$ is multidimensional, the region covered by $\varphi_{i1}^{(k)}$, $\varphi_{i2}^{(k)} \sim U(0, c)$ and $\varphi_{i1}^{(k)} \cdot (p_i^{(k)} - x_i^{(k)}) + \varphi_{i2}^{(k)} \cdot (p_l^{(k)} - x_i^{(k)})$ forms a hyperspace, while $\omega \cdot (x_i^{(k)} - x_i^{(k-1)})$ shifts this hyperspace by a step size, as illustrated in Fig. 6.3.

Thus, $\varphi_{i1}^{(k)}$ and $\varphi_{i2}^{(k)}$ determine the size of the feasible region R, while ω sets the step size for the movement of R. In other words, the hypercube defined by $\varphi_{i1}^{(k)}$ and $\varphi_{i2}^{(k)}$ moves within the search space depending on the magnitude of $|\omega|$ to explore the gbest. Therefore, $|\omega|$ influences the extent of the feasible region R and also controls the precision of the search within R. A larger $|\omega|$ expands the algorithm's search range in the solution space but reduces search precision, potentially missing the gbest g^*, and in some cases, might even cause the algorithm to diverge, failing to converge

Fig. 6.3 Feasible region R of a particle

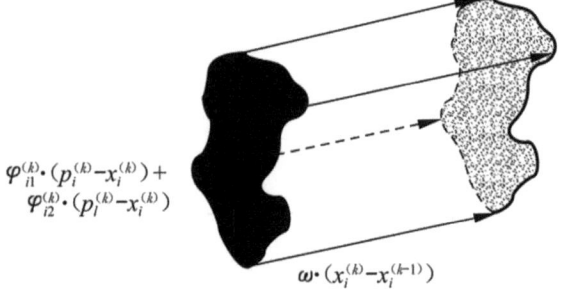

$$\varphi_{i1}^{(k)} \cdot (p_i^{(k)} - x_i^{(k)}) +$$
$$\varphi_{i2}^{(k)} \cdot (p_l^{(k)} - x_i^{(k)})$$

$$\omega \cdot (x_i^{(k)} - x_i^{(k-1)})$$

to the *gbest*. Conversely, a smaller $|\omega|$ enhances search precision but increases the search time and may lead to premature convergence to a *lbest*.

At time step $k + 1$, if $g^* \notin R$, then $\varepsilon_k = 0$. Therefore, $p(\xi_i^{(k+1)} \in M \,|\, \xi_i^{(k)} \notin M) = 0$. Conversely, if $g^* \in R$ at time step $k + 1$, then $\varepsilon_k = \varepsilon$. In this case, $p(\xi_i^{(k+1)} \in M \,|\, \xi_i^{(k)} \notin M)$ is inversely proportional to $|\omega|$. This means that the larger the step size determined by $|\omega|$, the smaller the probability $p(\xi_i^{(k+1)} \in M \,|\, \xi_i^{(k)} \notin M)$, thus increasing the likelihood that the algorithm will miss the *gbest*. On the other hand, when $|\omega|$ is smaller, $p(\xi_i^{(k+1)} \in M \,|\, \xi_i^{(k)} \notin M)$ is larger, indicating that, when the *gbest* g^* is within the feasible region R, a smaller $|\omega|$ allows the algorithm to take smaller steps, thus improving the search accuracy within the feasible region and finding the *gbest*. However, since $|\omega|$ controls the step size and hence the size of the feasible region R, it also affects whether the *gbest* g^* lies within R. A larger $|\omega|$ increases the probability that $g^* \in R$, while a smaller $|\omega|$ decreases this probability. Therefore, an appropriately chosen $|\omega|$ is crucial for achieving a favorable value for $p(\xi_i^{(k+1)} \in M \,|\, \xi_i^{(k)} \notin M)$.

6.3.3 Impact of the Acceleration Factor C

(1) When c is too small:

If the value of c is too small, even if the *gbest* position $p_l^{(k)}$ and the *pbest* position $p_i^{(k)}$ are far from the current position $x_i^{(k)}$ in the early stages of the algorithm, the small value of $c \cdot \left\| p_i^{(k)} - x_i^{(k)} \right\| \cdot c \cdot \left\| p_l^{(k)} - x_i^{(k)} \right\|$ results in a feasible region R that is too small. As iterations continue, $p_l^{(k)}$ and $p_i^{(k)}$ become closer to $x_i^{(k)}$, and $c \cdot \left\| p_i^{(k)} - x_i^{(k)} \right\| \cdot c \cdot \left\| p_l^{(k)} - x_i^{(k)} \right\|$ becomes even smaller. If $g^* \in R$ and $\varepsilon_k = \varepsilon > 0$, then $\lim_{k \to \infty} P(\varsigma^{(k+1)} \in \Gamma) = 1$, allowing the algorithm to quickly converge to the *gbest* g^*. However, if $g^* \notin R$ and $\varepsilon_k = 0$, then $\lim_{k \to \infty} P(\varsigma^{(k+1)} \in \Gamma) = 0$, causing the algorithm to converge rapidly to a single point (not the *gbest*). Therefore, if c is too small, the feasible region R becomes increasingly smaller, leading the swarm to quickly converge to a point and resulting in premature convergence of the algorithm.

(2) When c is too large:

Examining the feasible region R, if the inertia weight ω is known, even if the current positions $p_l^{(k)}, p_i^{(k)}$ are close to $x_i^{(k)}$, a large value of c will make $c \cdot ||p_i^{(k)} - x_i^{(k)}|| \cdot c \cdot ||p_l^{(k)} - x_i^{(k)}||$ excessively large. This results in the positions $p_l^{(k+1)}, p_i^{(k+1)}$ being far from $x_i^{(k+1)}$ at time step $k + 1$, which in turn causes $c \cdot ||p_i^{(k)} - x_i^{(k)}|| \cdot c \cdot ||p_l^{(k)} - x_i^{(k)}||$ to keep increasing. This makes it a monotonically increasing function. Consequently, as iterations proceed, the feasible region R continuously expands, leading the algorithm to diverge.

When PSO diverges, $c \cdot ||p_i^{(k)} - x_i^{(k)}|| \cdot c \cdot ||p_l^{(k)} - x_i^{(k)}||$ becomes a monotonically increasing function, indicating that $p(\xi_i^{(k+1)} \in M | \xi_i^{(k)} \notin M) = \frac{\varepsilon_k}{c \cdot ||p_i^{(k)} - x_i^{(k)}||} \cdot$ $\frac{\varepsilon_k}{c \cdot ||p_l^{(k)} - x_i^{(k)}||}$ decreases monotonically over time. That is, $P(\xi_i^{(k+1)} \in M | \xi_i^{(k)} \notin M)$ gets progressively smaller, and $\lim\limits_{k \to \infty} p(\xi_i^{(k+1)} \in M | \xi_i^{(k)} \notin M) \to 0$. Therefore:

$$\lim_{k \to \infty} \prod_{l=1}^{k} \prod_{i=1}^{m} \frac{\varepsilon_l^2}{c^2 \cdot ||p_i^l - x_i^l|| \cdot ||p_l^{(l)} - x_i^l||} \to 0$$

Thus, we have:

$$\lim_{k \to \infty} p(\varsigma^k \in \Gamma) = 1 - p(\varsigma^1 \notin \Gamma) \cdot \prod_{l=1}^{\infty} \prod_{i=1}^{m} \left(1 - \frac{\varepsilon_l^2}{c^2 \cdot ||p_i^l - x_i^l|| \cdot ||p_l^{(l)} - x_i^l||} \right) = 0$$

This suggests that when the PSO algorithm diverges, even if the feasible region R is large enough to contain the *gbest* $g^* \in R$, the algorithm may still fail to locate g^*. Therefore, even when $g^* \in R$, it does not guarantee that the algorithm will find the optimal solution g^*. It is important to note that when $g^* \in R$, the range of R should be appropriately bounded, that is:

$$p(\xi_i^{(k+1)} \notin M | \xi_i^{(k)} \notin M) = 1 - \frac{\varepsilon_k}{c \cdot ||p_i^{(k)} - x_i^{(k)}||} \cdot \frac{\varepsilon_k}{c \cdot ||p_l^{(k)} - x_i^{(k)}||} < 1$$

Moreover, $p(\xi_i^{(k+1)} \in M | \xi_i^{(k)} \notin M) = \frac{\varepsilon_k}{c \cdot ||p_i^{(k)} - x_i^{(k)}||} \cdot \frac{\varepsilon_k}{c \cdot ||p_l^{(k)} - x_i^{(k)}||}$ is not a monotonically decreasing function over time. Only when the iterations tend to infinity can we have $\lim\limits_{k \to \infty} p(\varsigma^{(k)} \in M) = 1$.

6.4 The Standard PSO Algorithm Has a Certain Probability of Finding the Gbest

Due to the inherent randomness in the parameter selection of the standard PSO, it cannot guarantee that the *gbest* g^* will always be within the feasible region R during every iteration, i.e., $\varepsilon_k \geq 0$, thus:

$$p(\xi_i^{(k+1)} \in M | \xi_i^{(k)} \notin M)$$
$$= \frac{\varepsilon_k}{|\omega| \cdot ||x_i^{(k)} - x_i^{(k-1)}||} \cdot \frac{\varepsilon_k}{c \cdot ||p_i^{(k)} - x_i^{(k)}||} \cdot \frac{\varepsilon_k}{c \cdot ||p_l^{(k)} - x_i^{(k)}||} \leq 1 \qquad (6.32)$$

This leads to $0 \leq p(\varsigma^\infty \in M) \leq 1$, meaning that the likelihood of the standard PSO finding the *gbest* is probabilistic in nature.

During the optimization process of the PSO algorithm, if at certain time steps the *gbest* $g^* \in R$, and the selection of algorithm parameters ω and c ensures that the algorithm does not diverge, then there exists some $j \in \{1, 2, \cdots, n\}$ such that

$$p(\xi_i^{(kj+1)} \in M \,|\, \xi_i^{(kj)} \notin M)$$

$$= \frac{\varepsilon}{|\omega| \cdot ||x_i^{(kj)} - x_i^{(kj-1)}||} \cdot \frac{\varepsilon}{c \cdot ||p_i^{(kj)} - x_i^{(kj)}||} \cdot \frac{\varepsilon}{c \cdot ||p_l^{(kj)} - x_i^{(kj)}||} > \delta$$

At this point, we have:

$$1 - \frac{\varepsilon}{|\omega| \cdot ||x_i^{(kj)} - x_i^{(kj-1)}||} \cdot \frac{\varepsilon}{c \cdot ||p_i^{(kj)} - x_i^{(kj)}||} \cdot \frac{\varepsilon}{c \cdot ||p_l^{(kj)} - x_i^{(kj)}||} < 1 - \delta$$

$$\prod_{j=1}^{\infty} (1 - \frac{\varepsilon}{|\omega| \cdot ||x_i^{(kj)} - x_i^{(kj-1)}||} \cdot \frac{\varepsilon}{c \cdot ||p_i^{(kj)} - x_i^{(kj)}||} \cdot \frac{\varepsilon}{c \cdot ||p_l^{(kj)} - x_i^{(kj)}||}) = 0$$

Thus, as $k \to \infty$, there will be n iterations where $g^* \in R$, and:

$$p(\varsigma^{(k+1)} \in \Gamma)$$

$$= 1 - p(\varsigma^{(1)} \notin \Gamma)$$

$$\cdot \prod_{l=1}^{k} \prod_{i=1}^{m} \left(1 - \frac{\varepsilon_k}{|\omega| \cdot ||x_i^{(k)} - x_i^{(k-1)}||} \cdot \frac{\varepsilon_k}{c \cdot ||p_i^{(k)} - x_i^{(k)}||} \cdot \frac{\varepsilon_k}{c \cdot ||p_l^{(k)} - x_i^{(k)}||}\right)$$

$$= 1 - p(\varsigma^{(1)} \notin \Gamma)$$

$$\prod_{j=1}^{n} \prod_{i=1}^{m} \left(1 - \frac{\varepsilon}{|\omega| \cdot ||x_i^{(kj)} - x_i^{(kj-1)}||} \cdot \frac{\varepsilon}{c \cdot ||p_i^{(kj)} - x_i^{(kj)}||} \cdot \frac{\varepsilon}{c \cdot ||p_l^{(kj)} - x_i^{(kj)}||}\right)$$

As $n \to \infty$, $\lim_{n \to \infty} p(\varsigma^{(k+1)} \in \Gamma) = 1$, causing the algorithm to converge to the *gbest*. However, in practice, the PSO algorithm cannot guarantee $n \to \infty$, resulting in $0 < p(\varsigma^{(k+1)} \in \Gamma) < 1$. That is to say, the standard PSO has a finite probability of finding the *gbest*.

6.5 Summary

This part defines particle and swarm state sequences based on the difference model of PSO, calculating transition probabilities of particles by leveraging the stochastic nature of the acceleration factors rather than merely treating the particle's feasible region as a hyperspace. Using the total probability formula, we first compute the probability for the swarm to transition to a non-optimal state, and from this, derive the probability of transitioning to the optimal state rather than simply calculating the probability of state transition for the swarm. Based on these results, we analyze the impact of acceleration factors on issues such as premature convergence and divergence. It is found that the parameter settings in the standard PSO cannot guarantee that the *gbest* position will always lie within the feasible region R. As a result, the standard PSO cannot ensure it will always find the *gbest*, nor can it guarantee it will fail to do so. However, careful selection of parameters can increase the likelihood that the optimal position falls within the feasible region, with an appropriately balanced range for R, thereby enhancing the probability of the algorithm converging to the *gbest* position.

References

1. Jin X-L, Ma L-H, Wu T-J et al (2007) Convergence analysis of the particle swarm optimization based on stochastic processes. Acta Automatica Sinica 33(12):1263–1268
2. Yuan D-L, Chen Q (2009) Particle swarm optimization algorithm based on Markov model and its stochastic process analysis. Comput Eng Appl 45(31):49–52
3. Li N (2006) Analysis and application of particle swarm optimization. Huazhong University of Science and Technology, Wuhan
4. Ren Z-H, Wang J, Gao Y-L (2011) The global convergence analysis of particle swarm optimization algorithm based on Markov chain. Control Theory Appl 28(4):462–466
5. Cai Z, Huang H, Zheng Z et al (2009) Convergence improvement of particle swarm optimization based on the expanding attaining-state set. J Huazhong Univ Sci Technol 37(6):44–47
6. Poli R, Langdon WB (2007) Markov chain models of bare-bones particle swarm optimizers. In: Proceedings of the 9th annual genetic and evolutionary computation conference, GECCO 2007, July 7–11, 2007, London, United Kingdom. Association for Computing Machinery

Chapter 7
Single-Objective PSO

Generally, the standard PSO is effective for addressing optimization problems. However, for more specific or complex problems, many researchers have developed various enhanced algorithms. These PSO variants primarily aim to improve solution accuracy, swarm diversity, and convergence speed. Moreover, certain variants have been applied to multi-objective optimization, dynamic optimization, constrained optimization, and dynamic multi-objective optimization problems. This chapter focuses on reviewing improved PSO algorithms designed for continuous, unconstrained, static, single-objective optimization problems.

7.1 Topology-Based Improved PSO

In PSO, the social network structure describes the relationships and interactions patterns between particles within the swarm. This structure governs information propagation among the particles and directly affects the optimization performance and convergence behavior. The standard PSO typically employs a global star topology for information exchange, where the optimal particle is selected from the entire swarm. In local-best problems, this strategy can lead the swarm to converge too quickly to a local optima, making escape difficult. Consequently, based on the algorithm characteristics, researchers have proposed alternative topologies, such as ring structures and Von Neumann structures.

The core concept of the ring-structured PSO is to replace the gbest particle with a lbest one, i.e., the lbest for each particle is determined by its immediate neighbors. A commonly used ring structure is illustrated in Fig. 7.1.

In the static neighborhood structure, the ring-structured PSO algorithm determines the neighborhood of each particle based on its index within the swarm. For the number of neighbors $k = 2$, the direct neighbors of a particle with index i are the particles with

© Beijing Institute of Technology Press 2025
F. Pan et al., *Particle Swarm Optimizer and Multi-Objective Optimization*,
https://doi.org/10.1007/978-981-95-3381-7_7

Fig. 7.1 The ring structure

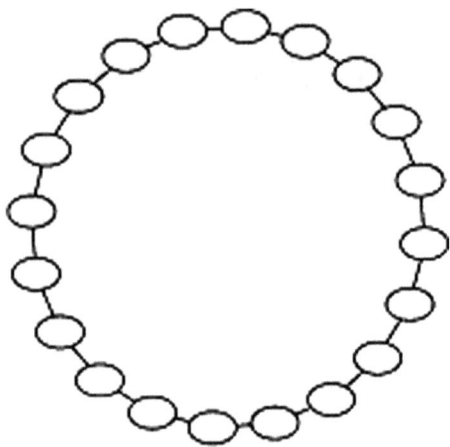

indices $(i-1)$ and $(i + 1)$. The method for selecting the *lbest* in the ring-structured PSO is shown as follows:

> *For i=1,..., m do*
> > *If* $1<i<m$
> > > $Lbest=\{x_s \mid f(x_s)=Min(f(x_{i-1}), f(x_i), f(x_{i+1}))\}$
> > *Elseif i*==1
> > > $Lbest=\{x_s \mid f(x_s)=Min(f(x_m), f(x_i), f(x_{i+1}))\}$
> > *Elseif i*==*m*
> > > $Lbest=\{x_s \mid f(x_s)=Min(f(x_{i-1}), f(x_i), f(x_1))\}$
> > *End*
>
> *end*

In the dynamic neighborhood structure, particles do not have fixed indices But instead determine their neighbors according to a particular strategy. Suganthan [1] proposed a method for defining neighborhoods based on the *Euclidean* distance between particles. For the number of neighbors $k = 2$, the neighborhood of particle i consists of the two closest particles. The method for selecting the *lbest* is as follows:

> *For i=1,..., m do*
> > *For j=1,..., m do*
> > > $L_{ij}=\varepsilon(x_i,x_j)$
> > *End*
> > $x_{i1}=\{x_p \mid L_{is}=Min(L_{ij}), j = 1,...,m\}$
> > $x_{i2}=\{x_q \mid L_{is}=Min(L_{ij}), j = 1,..., p\text{-}1, p+1,..., m \}$
> > $Lbest=\{x_s \mid f(x_s)=Min(f(x_i), f(x_p), f(x_{iq}))\}$
>
> *end*

Although the dynamic neighborhood increases the computational complexity by requiring the calculation of *Euclidean* distances between all particles, it dynamically changes the particle neighborhood. This enhances swarm diversity, thereby improving the global search capabilities of the algorithm.

Beyond the topological structure with $k = 2$, Suganthan [1] combined the ring structure with the star structure. In this approach, the PSO algorithm starts with $k = 2$, and as the iterations progress, the number of neighbors gradually increases until reaching $k = n-1$, eventually covering the entire swarm.

For the expanding neighborhood model, a particle s is added to the neighborhood of particle i when the following condition is satisfied:

$$\frac{||x_i^k - x_s^k||}{d_{\max}} < \varepsilon$$
$$\varepsilon = \frac{3k + 0.6n}{n} \tag{7.1}$$

where d_{max} is the maximum distance between two particles, and n is the maximum number of iterations.

7.2 PSO Algorithms Based on Mathematical Models

The velocity and position update equations in the standard PSO is shown in Eq. (1. 1). Many researchers have focused on improving the basic mathematical model of PSO to address theoretical issues, thus leading to the development of various new PSO variants. Common examples include PSO with inertia weight, PSO with a constriction coefficient, and Bare Bones Particle Swarm (BBPS). These approaches are primarily used for theoretical analysis of PSO and aim to extend its scope of applications. A detailed discussion of these methods can be found in Sect. 2.3.

7.3 Hybrid PSO

PSO is widely favored by researchers for its computational simplicity and ease of implementation. Moreover, PSO can be easily combined with other optimization algorithms, resulting in hybrid algorithms that often exhibit better performance than the original. By drawing analogies between algorithms and biological individuals, the algorithms mechanisms are equated to genes, and the combined algorithms reflect "swarm evolution." Currently, no theoretical framework guarantees that the hybrid algorithms will necessarily produce a superior algorithm compared to the originals— just as in biological evolution. The key to studying the mechanisms fusion lies in

identifying the strengths and weaknesses of different algorithms based on optimization principles. There are two primary ways to the mechanisms fusion: one is to
integrate with pure mathematical theories, and the other is to merge with mechanisms from other methods for intelligent optimization. For example, the combination of PSO and the genetic algorithm (GA) has been successfully applied to address
complex problems such as dynamic clustering [2], automated data generation in software testing [3], reservoir permeability prediction using neural network training [4],
multi-satellite observation scheduling [5], closed-loop supply chain planning [6], and
cost optimization for photovoltaic power [7]. The integration of PSO and simulated
annealing has solved issues like cluster analysis [8], urban water demand forecasting
[9], support vector machine (SVM) optimization [10], dynamic spectrum allocation
in broadcast networks [11], dynamic and optimal power dispatch in power systems
[12], and train scheduling optimization [13]. Xin et al. [14] provided a comprehensive
review of hybrid algorithms combining differential evolution (DE) with PSO, demonstrating that efficiently merging PSO with DE enhances the algorithm's balance
between "exploration" and "exploitation." The hybridization of PSO and ant colony
optimization (ACO) has been applied to problems such as multi-attribute partner
selection in virtual enterprises [15], Turkey's energy demand forecasting [16], multi-
processor task scheduling [17], and planar truss optimization [18]. Figure 7.2 shows
all the current optimization algorithms, and existing algorithms through mechanism
fusion can be combined to form new algorithms.

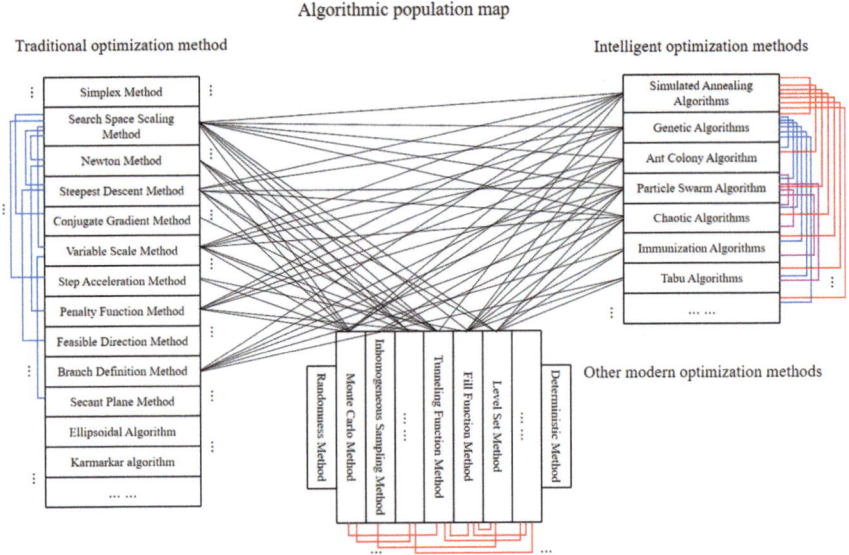

Fig. 7.2 Swarm dynamics of optimization algorithms

7.3.1 PSO Based on GA

Wang et al. [19] introduced a hybrid PSO that was improved by integrating genetic algorithm mutation operations. They employed a mutation operator to modify particle states, enhancing the swarm diversity and preventing local convergence as well as maintaining search capabilities.

Based on earlier research, Xie [20] incorporated operations of GA such as selection, crossover, and mutation into PSO, resulting in an improved PSO based on GA. The inertia weight was adjusted using a nonlinear decreasing approach to enhance the convergence of the algorithm.

The basic principle of the improved algorithm is as follows:

In Xie [20], a nonlinear inertia weight decreasing function is employed:

$$\omega = (\omega_{max} - \omega_{min})\left(\frac{k}{\text{MaxIteration}}\right)^2 + (\omega_{max} - \omega_{min})\frac{2k}{\text{MaxIteration}} + \omega_{min} \quad (7.2)$$

where MaxInteration represents the maximum number of iterations, t is the current number of iterations, ω_{max} represents the maximum inertia weight ω (typically ranging from 0.8 to 1.2), and ω_{min} represents the minimum inertia weight.

We now introduce and analyze the operators:

First, the selection operator is implemented. The operation schemes are chosen proportionally, with the appropriate ratio determined based on the specific problem. For example, the literature selects the top half of the particles, i.e., $m/2$ particles (where m is not the total number of particles in the swarm) based on the fitness evaluation function to directly enter the next generation. This ensures that the superior characteristics of the particle swarm are well preserved, while enhancing the search for better solutions.

For the crossover operator, the concept of crossover from GA is introduced, leading to the idea of crossover PSO. During each iteration, a certain number of particles are selected and placed into a pool. A random probability, unrelated to the fitness of the function, is assigned to the selected particles, referred to as the crossover probability. Based on this crossover probability, the selected particles undergo crossover operations, generating an equal number of new particles to replace those from the previous generation while ensuring that the total particle count unchanged. The positions of the new particles are calculated through an weighted arithmetic average of the original particle positions. A similar crossover method is applied, where the top half of the particles, sorted by fitness, are selected for pairwise crossover. A random crossover probability, unrelated to fitness, is assigned to produce an equal number of particles, which replace the bottom half of the particles for the next generation. The positional crossover is performed using the following equation:

$$\begin{aligned} x_1(i+1) &= \text{rand} \cdot x_1(i) + (1 - \text{rand}) \cdot x_2(i) \\ x_2(i+1) &= \text{rand} \cdot x_2(i) + (1 - \text{rand}) \cdot x_1(i) \end{aligned} \quad (7.3)$$

where x_i ($i = 1, 2$)is the position variable in the D–dimensional search space, and $x_1(i)$ and $x_2(i)$ are the two particles involved in the crossover operation. *Rand* is a random variable uniformly distributed between $[0, 1]$ in the D-dimensional search space. The velocity update equation for the new generation of particles is as follows:

$$\begin{aligned} v_1(i+1) &= \frac{v_1(i) + v_2(i)}{|v_1(i) + v_2(i)|}|v_1(i)| \\ v_2(i+1) &= \frac{v_1(i) + v_2(i)}{|v_1(i) + v_2(i)|}|v_2(i)| \end{aligned} \tag{7.4}$$

where $v_1(i)$ and $v_2(i)$ are the velocities of the two particles undergoing the crossover operation. The new velocity obtained after crossover replaces the velocity of the previous generation, thereby completing the crossover process for both velocity and position.

The mutation operation involves randomly applying mutations across the entire particle swarm, followed by reinitialization. A mutation probability P_m is set and compared with a randomly generated mutation probability. If certain conditions are met (e.g., the set mutation probability is greater than the randomly generated probability), the mutation is applied to the particle. The expression for P_m is as follows:

$$P_m = 0.10 - \text{Nowsize} \cdot (0.01)/\text{Size} \tag{7.5}$$

where *Nowsize* represents the current index of the particles. This mutation process helps prevent the algorithm from premature convergence.

The flowchart of the PSO based on GA is illustrated in Fig. 7.3.

7.3.2 PSO Based on Simulated Annealing

In the standard PSO algorithm, particle velocity is restricted during iterations to prevent excessive jumps, but there are fewer constraints on particle positions. This can result in position changes, thus slowing the convergence. By incorporating the concept of simulated annealing into the PSO algorithm, position constraints are introduced, preventing particle deterioration and speeding up swarm convergence.

Simulated annealing is inspired by the physical process of solid-state annealing, where a solid is heated to a high temperature and then gradually cooled. During the heating phase, the internal temperature rises, accelerating molecular motion and increasing internal energy. As cooling occurs, the particles stabilize, and by the time the solid reaches room temperature, its internal energy is minimized. According to the *Metropolis* criterion, the probability of a particle being in a stable state at temperature T is given by $e^{(-\Delta E/kT)}$, where E represents the internal energy at temperature

Fig. 7.3 PSO based on GA

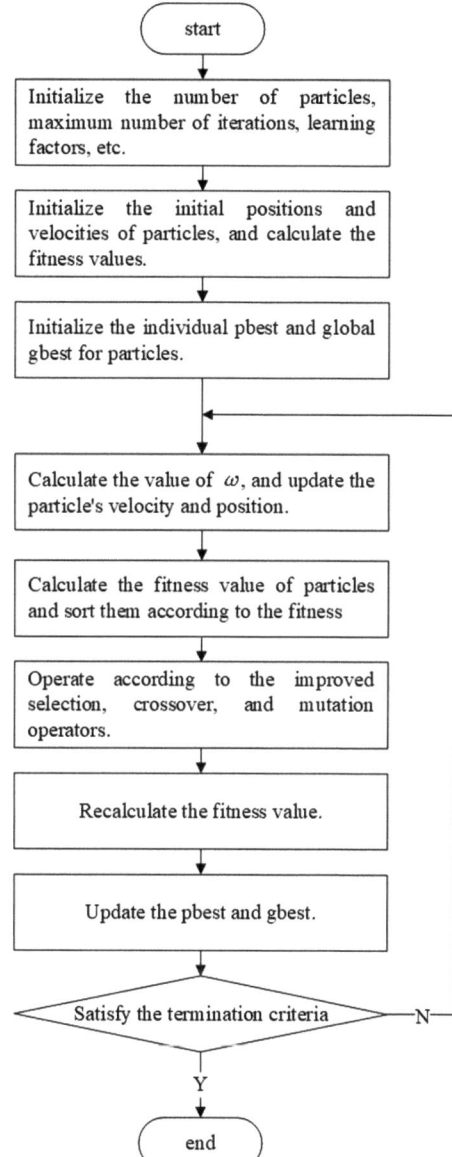

T, ΔE represents the energy change, and k represents the *Boltzman* constant. In optimization, E is set as the objective function f, and temperature T is transformed into the control parameter t. The evolution process works as follows: starts with an initial solution i and an initial control parameter t, the algorithm iteratively performs the following steps for the solution 'generation new solutions-calculation the difference from objective- either accepting or rejecting the new solution'. Over time, the value of t is gradually reduced. The final solution at the end of the algorithm is the desired result. This approach is a heuristic, stochastic search method based on Monte Carlo iterations. The annealing process is governed by a cooling schedule, which controls the initial value of t, the decay factor Δt, the number of iterations L for each value of t, and the stopping condition S.

The improvement strategies can be generally divided into two approaches: the first involves introducing the simulated annealing process into the iterations of the particle swarm, embedding the annealing mechanism within the swarm so that it focuses solely on particle generation and updating. The second approach leverages the core concept of simulated annealing without relying on temperature T as a variable. Instead, other parameters, such as iteration count or alternative criteria, are set to determine the acceptance criterion, preventing degradation of the objective function.

In Xie [20], the second approach is used, where the fitness increment of new particles relative to the original ones is employed to define the acceptance criterion. If $\Delta E \leq 0$, the particle swarm undergoes mutation-based updating. Conversely, if $\Delta E < e$, where e represents an upper limit that allows particle variation, the particle swarm is updated. Otherwise, the update is rejected. Since PSO is prone to getting trapped in *lbest*, $exp(-\Delta t'/T)$ is not used. During the later stages of evolution, if the swarm falls into a *lbest* and the probability of accepting new solutions $exp(-\Delta t'/T)$ becomes low, the particles cannot escape the *lbest*, reducing convergence performance. Employing ΔE and e to control particle updates, the particles gain greater flexibility, enabling escape from lbest, thus improving the algorithm's convergence performance.

The process for PSO based on simulated annealing is as follows:

Step 1: Initialize the velocities and positions of the particles, and set the value of e with allowed variation;

Step 2: Generate new positions for each particle based on their current positions and velocities;

Step 3: If the number of iterations is below the specified limit, update the velocities of the particles, constraining them within a larger and predefined range;

Step 4: Calculate ΔE. If $\Delta E \leq 0$, accept the new particle;

Step 5: If $\Delta E < e$ (where e is the upper bound of the allowed variation), accept the new particle position; otherwise, reject it;

Step 6: Continue iterating. If a particle's fitness exceeds its historical optimum, replace it with the current best and identify both the *gbest* and *pbest*;

Step 7: If the termination criteria are met, end the iterations and output the optimal solution.

7.3.3 Chaotic Particle Swarm Optimization (CPSO) Based on Chaotic Optimization Principles

The inherent randomness in the initialization and evolution of the basic PSO algorithm may lead to inconsistent particle quality, negatively affecting convergence performance and making the algorithm prone to premature convergence at *lbest*. In order to address these issues, *Tang* et al. [21] used the ergodic properties of chaotic motion and proposed Chaotic Particle Swarm Optimization (CPSO) based on chaotic optimization principles. This algorithm not only improves the PSO's ability to escape the *lbest* but also enhances convergence speed and accuracy.

The core ideas of this algorithm are reflected in two aspects:

(1) Chaotic sequences are used to initialize particle positions and velocities. This approach maintains the randomness intrinsic to PSO's initialization while utilizing chaos to increase swarm diversity and enhance the ergodic of the particle search. After generating a large number of initial swarms, the best one is selected as the starting point.

(2) Chaotic sequences are generated based on the best position found by the swarm up to the current iteration. The optimal particle position from the chaotic sequence replaces the position of a particle in the current swarm. By incorporating chaotic sequence, the algorithm generates multiple neighborhood points around *lbest* during iterations, allowing inert particles to escape local minima and facilitating a faster search for the global optimal solution.

The logistic equation below represents a typical chaotic system:

$$z_{n+1} = \mu z_n(1 - z_n), n = 0, 1, 2, ...Z_{n+1}$$

where μ is a control parameter set to $\mu = 4$, and $0 \leq z_0 \leq 1$, indicating that the system is in a fully chaotic state. From any initial value $z_0 \in [0, 1]$, a deterministic time series can be generated as $z_1, z_2, z_3...$

The steps for CPSO are as follows:

Step 1: Initialize the maximum allowed number of iterations or the fitness error threshold, along with the parameters related to the CPSO algorithm (inertia weight and learning factors).

Step 2: Chaotic initialization of particle positions and velocities. Randomly generate an n-th dimensional vector with each component in the range of 0 ~ 1, $z_1 = (z_{11}, z_{12},..., z_{1n})$, where n is the number of variables in the objective function. Based on the chaotic system, generate N vectors $z_1, z_2,..., z_N$; map the components of z_i to their corresponding variable ranges; calculate the fitness of the particle swarm, and select M of the best-performing solutions from the initial N swarms as the initial solutions, randomly generating M initial velocities.

Step 3: If the particle's fitness exceeds its *pBest*, update *pBest* to the new position.

Step 4: If the particle's fitness exceeds the *gBest*, update *gBest* to the new position.

Step 5: Update the particle's velocity and position.

Step 6: Perform chaotic optimization on the optimal position $p_g = (p_{g1}, p_{g2,...,} p_{gD})$. Map $p_{gi}(i = 1, 2, ... D)$ to the domain of the *Logistic* equation [0, 1] using the transformation $z_i = (p_{gi}-a_i)/(b_i-a_i))$ $(i = 1, 2, ..., D)$. Then, iterate using the *Logistic* equation to generate a chaotic sequence $z_i^{(m)}$ $(m = 1, 2, ...)$, and map the generated chaotic sequence back to the original solution space using the inverse mapping $p_{gi}(m) = a_i + (b_i-a_i))z_i^{(m)}$ to obtain $P_g^{(m)} = (P_{g1}^{(m)}, P_{g2}^{(m)},...,P_{gD}^{(m)})$, $(m = 1,2,...)$. In the original solution space, calculate the fitness for each feasible solution $P_g^{(m)}$ $(m = 1, 2,...)$ to identify the best-performing feasible solution P^*.

Step7: Replace the position of any particle in the current swarm with P^*.

Step8: If the stopping conditions are met, terminate the search and output the gbest position; otherwise, return to Step 3.

Tang [21] evaluates the performance of the CPSO optimization algorithm using four test functions. The results indicate that CPSO exhibits the lowest optimal fitness and average optimal fitness, making it more precise and stable compared to PSO and GA. This advantage is particularly pronounced in cases with higher dimensionality and rapid changes in function value, where PSO performance significantly declines. From the average results, GA demonstrates greater variability, with a larger average fitness and standard deviation, and becomes increasingly unstable as dimensionality increases. Based on these results, CPSO proves to be an effective tool for global optimization of complex functions. For high-dimensional functions with numerous local extrema, CPSO can achieve a certain level of accuracy in finding the global minimum at a lower computational cost.

7.3.4 Swarm Intelligence Algorithm Based on PSO and Shuffled Frog Leaping

In order to address premature convergence issues in both PSO and the shuffled frog leaping algorithm (SFLA), Sun et al. [22] proposes a novel swarm intelligence algorithm that integrates both methods. The entire swarm is divided into equal-sized groups: one representing the frog population and the other the particle swarm. During their independent evolution, a dynamic information exchange strategy is employed between the two groups. The best fitness values of the frog population and particle swarm are compared: if the frog group has evolved better, the worst-performing individuals in each frog subgroup are replaced by some of the better-performing individuals from the particle group. Conversely, if the particle group shows better performance, some of its stronger individuals replace the best individuals in each frog subgroup. Additionally, a collaboration mechanism is introduced between the two groups. In order to prevent the particle group from suffering premature convergence, which would reduce the effectiveness of the information exchange, random perturbations are applied to the *pbest* positions of all individuals at appropriate times.

The steps of the algorithm are as follows:

Step l: Initialize the parameters of the particle swarm and frog population. Record the *gbest* and *pbest* for the particle swarm, the best and worst frogs in each frog subgroup, and the best frog in the entire frog population.

Step 2: Set the inertia weight ω to decrease linearly from 1.8 to 0.9.

Step 3: Sort the fitness values of the frog population and divide them into m subgroups. Similarly, sort the *pbest* of the particle swarm and save the top m into $P_B[j]$. In addition, store the *pbest* and *pworst* from each frog subgroup into $P_b[j]$ and $P_w[j]$, respectively.

Step 4: Compare the best fitness GX obtained by the particle swarm and the best fitness value GY obtained by the frog population, and execute the information exchange strategy. If GX is better, further compare the fitness values of $P_B[m-j+1]$ from the particle swarm with $P_b[j]$ from the frog population. If the fitness of $P_B[m-j+1]$ is better, replace the position of $P_b[j]$ with $P_B[m-j+1]$; otherwise, do not replace. Then, perform a local search based on local search equations for the frogs. If GY is better, further compare the fitness values of $P_w[j]$ from the frog population with $P_B[j]$ from the particle swarm. If the fitness of $P_w[j]$ is worse, replace the position of $P_B[m-j+1]$ with $P_w[j]$; otherwise, do not replace. Then, perform a search using the update equations for velocities and positions in PSO.

Step 5: Check whether the number of times GX remaining unchanged has reached the specified threshold. If it does, apply random perturbations to the *pbest* of the particle swarm as follows (where p_j is the *pbest* of the j-th particle, and r is a random number in the range [0, 1]). Otherwise, do not perturb.

$$p_j^{(k+1)} = p_j^{(k)} + r \cdot p_j^{(k)}$$

Step 6: Output the better solution from the two groups as the result. Check the termination condition of the iteration. If the condition is met, stop; otherwise, return to Step 2.

In Sun et al. [22], the performance of the algorithm was compared using four test functions. Experimental results show that the new algorithm demonstrates excellent performance in global search and effectively overcomes premature convergence compared with PSO and SFLA, which tend to fall into local optima. Additionally, the convergence speed is significantly improved, and the algorithm shows higher stability when solving high-dimensional problems.

7.4 Multi-swarm PSO

The multi-swarm PSO aims to maintain diversity, enabling the algorithm to escape lbest and enhance its optimization performance. In a multi-swarm PSO, each subswarm operates independently according to different algorithmic rules. Simultaneously, there is a cooperative aspect: the velocity update of each particle is updated not only by *pbest* position p_i found by individuals and the *gbest* position p_l in that swarm, but also by the overall best position p_l' relative to the best positions of all sub-swarms. In addition to the existing "cognitive" and "social" components of improved particle swarm, this approach introduces a new component, tentatively referred to as the "supra-social" component, which incorporates the concepts of cooperation and sharing into the multi-swarm algorithm. This reintroduces the idea of collaboration within the multi-swarm PSO. Given the characteristics and relationships between PSO and multi swarm, the concepts of expansion mutation and perturbation are also integrated. When a particle is at the best position of its swarm, it can easily become trapped in *lbest*. Thus it is necessary to apply expansion mutation to facilitate its escape. While the multi-swarm is diverse and exhibits independent characteristics, it still operates under specific rules, which can lead to stagnation. Therefore, after a certain number of iterations, a perturbation is applied to the entire swarm.

7.4.1 Tabu Search-Based Bi-Group Particle Swarm Optimization (TSBBPSO)

Tabu search [23] (TS) is a global neighborhood search that simulates the optimization characteristics of human memory. TS employs a local neighborhood search along with specific tabu criteria to avoid cyclical searches and utilizes tabu relaxation to release certain restricted superior states, thereby ensuring effective exploration and ultimately achieving global optimization. UP to date, TS has been widely applied in various fields, including combinatorial optimization, production scheduling, machine learning, and neural networks, achieving notable success. In recent years, it has also garnered increased research interest for its application in function global optimization, indicating a promising trend for future development.

Liu [24] introduces the tabu search based bi-group particle swarm optimization (TSBBPSO). The core concept is to divide the particle swarm into two sub-swarms, S_1 and S_2, of different sizes. S_1 uses a continuously decreasing inertia weight for search, while S_2 employs a larger inertia weight. As the evolution progresses, the number of particles in each sub-swarm is dynamically adjusted: in the early stages of evolution in order to emphasize global optimization capabilities, S_2 contains more particles to enhance the swarm's global search ability. Conversely, in the later stages, in order to require stronger local search capabilities, S_1 should have a greater number of particles to improve its local search performance. The number of particles in S_1 and S_2 is determined by the following equations:

$$\Delta m = \left\lfloor 2 \cdot \frac{m}{3} \cdot (1 - \frac{k}{\text{MaxIteration}}) \right\rfloor$$
$$S_1 = \left\lfloor \frac{5}{6}m - \Delta m \right\rfloor \tag{7.6}$$
$$S_2 = m - S_1$$

where m represents the number of particles, *MaxIteration* represents the maximum number of iterations, and k represents the current number of iterations.

During each generation's sub-swarms regrouping phase, an elitist strategy is employed: particles with lower fitness values are added to S_2 to enhance global search capability, while particles with higher fitness values are assigned to S_1 to strengthen local search capability.

After several iterations, the particles will have identified initial solutions that are relatively good. At this stage, TS is introduced by calculating the fitness values of the particles and adding the particle with the best fitness to the tabu list. With the *heart* table, a new particle is reinitialized to replace the current best particle. When a particle's tabu count in the tabu list reaches 0, it is removed from the list and replaces the worst-performing particle in the population. The use of the tabu list enhances the short-term memory of the swarm. While this may slightly slow down the algorithm's convergence speed, it effectively prevents the swarm from prematurely falling into *lbest* and increases the chances of finding the *gbest*.

Equation (7.7) is used to determine whether the current optimal solution belongs to this region.

$$(best_{ij} - heart_j)^2 - \varepsilon \qquad 1 \le j \le D \qquad (7.7)$$

If the result of Eq. (7.7) is less than 0, it indicates that the particle belongs to the region, and the tabu frequency of that region is incremented by 1. Otherwise, it does not belong. In this case, $best_{ij}$ represents the position of the current best particle, with the *heart* table recording the center point of each region, and ε is a small value.

When the tabu frequency of a region reaches a certain multiple of a number, a neighborhood search is performed in that region, and the *heart* table is updated accordingly. Flexibly using the tabu frequency table takes full advantage of focused search of the tabu algorithm and enhances the swarm's long-term memory. This enables continuous neighborhood searches in regions likely to contain the *gbest*. During these neighborhood searches, certain particles in the region are reinitialized, which functions like a mutation. However, this mutation is targeted, not random, which improves the swarm's ability to accurately find the optimal solution while avoiding the instability caused by a complete random reinitialization.

The steps of the algorithm are as follows:

Step 1: Set the parameters, such as the number of particles m, the initialization range, maximum number of iterations *MaxIteration*, and learning factors c_1 and c_2. Randomly generate the positions and velocities of m particles within the space of feasible solutions, forming the initial particle swarm. Also, initialize the tabu list, the tabu frequency table, etc.;

Step 2: Calculate the fitness values of all particles and initialize p_l and p_l;

Step 3: If the current number of iterations exceeds the maximum allowed iterations, the algorithm terminates. Otherwise, divide the original swarm S into two sub-swarms, S_1 and S_2, of different sizes according to Eq. (7.6);

Step 4: Apply a linearly decreasing inertia weight $\omega_1 = 0.8 - \frac{0.7298 \cdot k}{\text{MaxIteration}}$ to S_1 and a fixed inertia weight of $\omega_2 = 0.8$ to S_2. For both sub-swarms, compute the current velocities and positions using Eq. (1.6), calculate the current fitness values, and update p_i and p_l for each sub-swarm;

Step 5: Combine S, S_1, and S_2 into a temporary swarm S'. Sort S' by fitness, then select the top m particles to generate the new swarm;

Step 6: If the number of iterations is less than the minimum allowed iterations N_t required to initiate the tabu search, return to Step 3;

Step 7: If the tabu list is not empty, reduce the tabu count of each particle in the tabu list. If a particle's tabu count reaches 0, adds it to the swarm, replacing the particle with the worst fitness;

Step 8 Use Eq. (7.7) to determine the region to which p_g belongs, then add the particle to the tabu list. Afterward, regenerate a new particle to replace p_g using the *heart* table, and increase the tabu frequency for that region by one. Recalculate the fitness values of all particles and update p_g accordingly. If the tabu frequency for a region reaches a multiple of a specified integer, perform a neighborhood search and update the *heart* table. Then return to Step 4.

In Liu [24], this algorithm was used to optimize five representative test functions and compared with standard PSO and GA, demonstrating the effectiveness of TSBBPSO. Additionally, TSBBPSO was applied to the *Packing* Problem, successfully solving it and providing an effective method for addressing this type of problem.

7.4.2 Vertical Parameter Multi-swarm Particle Swarm Optimization (VPMSPSO)

The real-world optimization problems may exist multiple optima. In multimodal optimization, it's sometimes necessary to identify each of these optima. By treating each particle as an initial sub-swarm, dynamic clustering is employed during sub-swarm movement based on specific rules, eventually forming sub-swarms that correspond to the number of optima. The evolutionary process involves two components: particle updates and sub-swarm updates. Chang [25] proposed a vertical parameter multi-swarm particle swarm optimization (VPMSPSO) algorithm, and its operational rules can be summarized as follows:

For the sub-swarms initialization, the vertical parameters between sub-swarms are calculated, including swarm distance, proximity, minimum distance after n iterations of the swarm in generation k, and position degree. In the set of n-iterations proximity distance in generation k, any two sub-swarms with a proximity distance smaller than the threshold and the closest position degree are merged. This merging operation is the sub-swarm update. The calculation for the next generation of particles then begins. This process repeats until M (the number of optima) sub-swarms are found, each of which iterates to discover its own optimal solution.

The swarm distance between sub-swarms refers to the distance between the optimal points of each sub-swarm. The swarm distance between sub-swarms i and j in generation t is denoted as $\Delta sb_{i,j}(t)$, and it is defined as follows:

$$\Delta sb_{i,j}(t) = \left\| sb_i(t) - sb_j(t) \right\| \tag{7.8}$$

where $sb_i(t)$ represents the optimal point of sub-swarm i in generation t. The choice of using the optimal point as a measure of sub-swarm position is due to the significant oscillation in particle positions, while the optimal positions are relatively stable.

The proximity between sub-swarms refers to the distance ratio between two sub-swarms separated by n generations. The proximity between sub-swarms i and j in generation t after n iterations is denoted as $\pi_{i,j}^n(t)$, and is defined as:

$$\pi_{i,j}^n(t) = \begin{cases} \dfrac{\Delta sb_{i,j}^t - \Delta sb_{i,j}^{t+n}}{\Delta sb_{i,j}^t + \Delta sb_{i,j}^{t+n}}, & \Delta sb_{i,j}^t + \Delta sb_{i,j}^{t+n} \neq 0 \\ 0, & otherwise \end{cases} \tag{7.9}$$

Proximity reflects the degree and speed at which the two sub-swarms converge during the evolution. A higher proximity indicates that the two sub-swarms are converging more quickly and to a greater extent; conversely, a lower proximity signifies a slower and lesser degree of convergence. The absolute value of $\pi_{i,j}^n(t)$ is less than 1, with $\pi_{i,j}^n(t) > 0$ indicating that they are approaching each other, otherwise, it indicates that they are moving apart. Additionally, the proximity of sub-swarm i to all other sub-swarms in the entire swarm at generation t after n iterations is denoted as θ:

$$\theta = \{\pi_{i,j}^n(t) | j = 1, 2, ..., N\} \tag{7.10}$$

where N represents the number of sub-swarms at generation t. The proximity of sub-swarm i to itself is 0.

The distances between all sub-swarms at generation t form the set: $S(t) \cdot S(t) = \{\Delta sb_{i,j}(t) | (i,j) \in \mathbf{N}+^{N \times N}\}$. For $\pi_{i,j}^n(t) > 0$, it can be regarded as the proximity set $S_j(t)$, while for $\pi_{i,j}^n(t) \leq 0$, it can be regarded as the distancing set $S_y(t)$. Thus, it follows that $S(t) = S_y(t) \cup S_j(t)$.

The minimum and maximum distances between sub-swarms at generation t are denoted as Δ_{min}^t and Δ_{max}^t:

$$\begin{aligned} \Delta_{min}^t &= \{\Delta sb_{i,j} | \min(\Delta sb_{i,j} \in S(t))\}, \\ \Delta_{max}^t &= \{\Delta sb_{i,j} | \max(\Delta sb_{i,j} \in S(t))\}. \end{aligned} \tag{7.11}$$

The distance ratio between sub-swarms i and j at generation t is denoted as $\lambda_{i,j}^t$:

$$\lambda_{i,j}^t = \frac{\Delta sb_{i,j}^t}{\Delta_{max}^t} \tag{7.12}$$

This can be interpreted as normalizing the distance.

The position degree of sub-swarms i and j at generation t is denoted as $\beta_{i,j}^t$:
$\beta_{i,j}^t = 1 - \lambda_{i,j}^t$.

The values of $\lambda_{i,j}^t$ and $\beta_{i,j}^t$ both range between $(0, 1]$. A smaller $\lambda_{i,j}^t$ and a larger $\beta_{i,j}^t$ indicate that the two centers are closer. When $\beta_{i,j}^t = 0$, the distance between the two sub-swarms reaches the maximum value within the entire swarm. The position degree measures the relative positioning between two sub-swarms, while a sub-swarm's position degree relative to all other sub-swarms reflects the distribution of sub-swarms centered around it within the entire swarm, which highlights the diversity of the sub-swarm.pi.

The introduction of these vertical parameters quantifies the relationships between sub-swarms throughout the evolution, and sub-swarm evolution operates based on these vertical parameters. Thus, the evolution rules for sub-swarms are defined as follows:

Step 1: Initialize all particles as single-particle sub-swarms;
Step 2: Update particles using standard PSO or any improved variant;
Step 3: Merge two sub-swarms if their proximity exceeds the threshold θ_π and their distance ratio falls below the threshold θ_λ;
Step 4: Stop evolution when all sub-swarms cease evolving or when the maximum number of generations is reached.

The steps for VPMSPSO are as follows:

Step 1: Initialize the algorithm parameters, including inertia weight, learning factors, proximity threshold θ_π, distance ratio threshold θ_λ, error threshold ε, and the maximum number of iterations t_{max};
Step 2: Initialize the velocity and position of each particle in the swarm, treating each particle as an individual sub-swarm. Calculate the fitness value of each particle, and initialize the *pbest* $pb_i(0)$ and sub-swarm best $sb_i(0)$, along with the distance matrix Δ, proximity matrix $\pi_{i,j}(0)$, and distance ratio matrix $\lambda_{i,j}^0$;
Step 3: Update the velocity and position of the particle swarm;
Step 4: Calculate the fitness of each particle, and update the *pbest* $pb_i(t)$ and the sub-swarm best $sb_i(t)$;
Step 5: Update matrix Δ, and calculate $\pi_{i,j}^n(t)$ and $\lambda_{i,j}^t$ for the best particle in each sub-swarm;
Step 6: Compare $\pi_{i,j}^n(t)$ and $\lambda_{i,j}^t$ with the thresholds θ_π and θ_λ. If the conditions are met, merge the sub-swarms. Update the number of sub-swarms K and $sb_k(t)$;
Step 7: If $t \geq t_{max}$, terminate the process;
Step 8: If no further sub-swarm merging occurs and $|(sb_k(t-1) - sb_k(t))/sb_k(t)| < \varepsilon$, it indicates that the swarm has stopped evolving and merging;
Step 9: End the process.

Chang [25] conducted extensive numerical experiments on VPMSPSO. Results showed that this algorithm effectively prevents peak loss during optimization with multimodal functions, and it offers superior multimodel search performance compared to other algorithms.

7.4.3 Extension-Based Multi-particle Swarm Optimization (EMPSO)

Extenics [26] is a novel discipline pioneered by Chinese scholars, primarily aimed at resolving contradictions in societal issues. Since its inception, Extenics has developed a comprehensive theoretical framework and a range of practical applications. In recent years, research combining Extenics with other optimization algorithms has also seen progress, such as the development of the Extenics genetic algorithm, which opens new avenues for integrating Extenics with intelligent optimization.

In Zhang [27], the concepts of Extenics were incorporated into PSO, resulting in the development of a multi-swarm PSO based on Extenics. Compared to traditional PSO, this approach is more biologically inspired and closely mimics the real-world foraging behavior of bird flocks. It enables particles at different levels to adopt distinct evolutionary strategies and employs the idea of extension transformation to revive inactive particles for further exploration.

Extension-based multi-particle swarm optimization (EMPSO) effectively addresses the limitations of standard PSO. With Extenics' evaluation methods and an association function, particles are classified into different ranks according to their fitness values. Particles in each rank adopt different inertia weights and evolve in a hierarchical manner. In terms of the bird flock, after identifying the "smart bird", other birds adjust their flight patterns based on their distance from the "smart bird", with the extent of their motion determined by proximity to the smart bird rather than the number of iterations. The concept of EMPSO resembles the foraging behavior of bird flocks in the real world. Even at the beginning of for aging, some birds will be near the "smart" birds, while as foraging finishes, there will still be birds that are far away. It is unreasonable for all birds to adopt the same motion patterns, whether at the beginning or near the end of foraging, as this would impede the flock's ability to find food sources more quickly and efficiently.

In addition, EMPSO uses the principle of extension transformation from Extenics to address particles with extremely low fitness values. These particles undergo an extension transformation, converting them into particles that contribute positively to the swarm, while retaining their previously attained optimal positions to ensure the diversity of the swarm is not affected.

EMPSO consists of four main components: matter-element representation, particle stratification, replacement transformation, and reassignment.

Table 7.1 Relationship between inertia weight and the rank of particles treated as matter-elements

Particle level	First	Second	Third	Fourth
$K(x)$	>0	$[-0.5, 0]$	$(-1, -0.5)$	≤ -1
ω	0.7(1-*gbest*/fitness)	0.7	Permutation exchange	Reinitialize

① Matter-element representation of particles

EMPSO treats each particle as a matter element, considering the particle's current position, personal best (*pbest*), fitness value, and inertia weight as characteristics of the particle's matter element. Based on the quality of the particle's fitness value, an attribute for particle ranking is added, categorizing the particles into different ranks. Particles at different ranks are assigned different values of inertia weight to achieve layered evolution of the particles.

The matter-element model of the particle is:

$$R = \begin{pmatrix} \text{Particle} & \text{Current position} & x^k \\ & \text{Personal best (pbest)} & p^k \\ & \text{Fitness value} & \textit{fitness} \\ & \text{Belonging rank} & \textit{Rank} \\ & \text{Inertia weight} & \omega \end{pmatrix} \tag{7.13}$$

② Particle stratification

The entire swarm is considered an extension set, with an association function[1] for calculation. Particles are divided into different ranks based on the values of the association function, allowing particles at each rank to adopt different values of inertia weight. The relationship between a particle rank and the corresponding value of the association function is illustrated in Table 7.1.

③ Permutation exchange

For particles located at the third rank, a permutation exchange is applied: $T(R_i) = R_{best}$, i.e.,

$$_RT_BB = (T_RR, c, _RT_vv) = \begin{pmatrix} T_R \text{ Particle Current position} & _RT_vx^k \\ \text{Personal best} & p^k \\ \text{Fitness value} & _RT_v\textit{fitness} \\ \text{Belonging rank} & _RT_v3 \\ \text{Inertia weight} & _RT_v\omega_3 \end{pmatrix} \tag{7.14}$$

---

[1] For details on the association function, refer to Zhang [27] by Zhang Hongbo. Research on Extension-based multi-particle swarm optimization Based on Extenics [D]; Beijing University of Chemical Technology, 2010. Equations (7.3)–(7.12).

④ Reinitialize

For particles at the fourth level, their position coordinates are reinitialized. The specific mathematical model for the algorithm is as follows:

Based on the dimensionality of the solution, the dimensionality of the particles is set, assuming the solution is D-dimensional. In a D-dimensional space, swarm i is randomly generated, and each swarm contains m particles. The positions, velocities, and *pbest* vectors of the particles are defined as follows:

$$X_{ij} = (x_{ij}^1, \cdots, x_{ij}^D) \quad i = 1, \cdots, l; j = 1, \cdots, m$$
$$V_{ij} = (v_{ij}^1, \cdots, v_{ij}^D) \tag{7.15}$$
$$P_{ij} = (p_{ij}^1, \cdots, p_{ij}^D)$$

where x_{ij}^q represents the position of the k-th dimension for the j-th particle in the i-th swarm. The *lbest* found within the i-th swarm is represented as $l_i = (l_i^1, \cdots, l_i^D)$, while the *gbest* found across the entire swarm is denoted as $G = (g^1, \cdots, g^D)$. The update equations for the positions and velocities of the particles are as follows:

$$V_{ij}^{(k+1)} = \omega_t \cdot V_{ij}^{(k)} + c_1 \cdot r_1 \cdot (P_{ij}^{(k)} - X_{ij}^{(k)}) + c_2 \cdot r_2 \cdot (L_{lj}^{(k)} - X_{ij}^{(k)})$$
$$+ c_3 \cdot r_3 \cdot (G^{(k)} - X_{ij}^{(k)})$$
$$X_{ij}^{(k+1)} = X_{ij}^{(k)} + V_{ij}^{(k+1)} \tag{7.16}$$

The steps for the *EMPSO* are as follows:

Step 1: Construct matter-elements for particles. Represent each particle and its associated attributes as a multi-dimensional matter-element, where the particle's current position, *pbest*, fitness function, rank, and inertia weight are defined as features. The form of the matter-element is as shown in Eq. (7.13);

Step 2: Initialization. In a D-dimensional space, initialize a set of particle matter-elements containing l sub-swarms, with each sub-swarm comprising n particles. Randomly initialize the position and velocity attributes of the matter-elements within the defined range. Set the upper and lower bounds for position and velocity updates (X_{min}, X_{max}) and (V_{min}, V_{max}). Initialize the *pbest* to the particle's initial position and set the *lbest* and *gbest* to the initial position of the first particle;

Step 3: Compute and update the fitness values of each matter-element. Calculate the current fitness of each particle using a fitness function and update the fitness attribute accordingly. The fitness function is determined by the objective function that needs to be optimized;

Step 4: Best position update: If the particle's current fitness exceeds its previous best, update the *pbest* position to its current position. If the current fitness surpasses the swarm's best fitness, update the swarm's *lbest* position to the particle's current position. Similarly, if the current fitness is better than the global fitness, update the *gbest* position to the particle's current position;

Step 5: Calculate the correlation degree of the matter-elements;
Step 6: Classify the matter-elements. Based on their correlation degree, categorize the matter-elements following the rules provided in Table7.1. Update the rank of each matter-element and adjust the inertia weight for particles in different ranks;
Step 7: Replacement transformation. Apply replacement transformation to matter-elements in the third rank, as defined in Eq. (7.14);
Step 8: Reinitialize. Reinitialize the matter-elements in the fourth rank;
Step 9: Positions and velocities update. Use Eq. (7.16) to iteratively update the position and velocity vectors of the matter-elements;
Step 10: Termination. If the termination condition is met, stop the iteration and output the optimal solution. Otherwise, return to Step 3.

In Zhang [27], four test functions were used to evaluate the performance of EMPSO. Comparisons with other multi-swarm PSO algorithms demonstrated the superior efficiency of EMPSO.

7.4.4 Adaptive Two-Swarm Particle Swarm Optimization (ATSPSO)

The social cognition aspect of PSO involves the selection of neighborhoods. When all particles in the swarm are treated as neighborhood members, the algorithm operates in its *gbest* version. When only a subset of members forms the neighborhood, it becomes the *lbest* version. In the *gbest* version, due to the larger scope of information sharing, the convergence speed is relatively fast. However, if the *gbest* position, p_g, is the *lbest*, the algorithm may experience premature convergence. The *lbest* version, on the other hand, has stronger global search capabilities, as multiple neighborhoods may each contain potential optimal seeds, which effectively helps avoid getting trapped in *lbest*. However, it converges more slowly.

Pan [28] combines these two versions and proposes an algorithm known as adaptive two-swarm particle swarm optimization (ATSPSO). In this approach, the swarm is divided into two sub-swarms: one employs the *gbest* version of the algorithm, while the other uses the *lbest* version. Typically, dividing the swarm requires increasing swarm size to ensure that each sub-swarm has sufficient search capability, but this can reduce the efficiency of the algorithm. To address this issue, the algorithm introduces an adaptive mechanism that adjusts the sizes of the two sub-swarms based on the characteristics of the problem and the convergence behavior observed during iterations. This improves both the search capability and the efficiency of the algorithm.

The basic principle of adaptive adjustment is as follows: when the particles in the *gbest* sub-swarm begin to converge, some particles from the *gbest* sub-swarm are moved to the *lbest* sub-swarm, thereby increasing its size and enhancing its global search capability, allowing the swarm to escape *lbest*. When the particles in the *gbest* sub-swarm are more dispersed, some particles from the *lbest* sub-swarm are transferred to the *gbest* sub-swarm, improving its local search ability and increasing the overall convergence speed and performance of the swarm.

Let N represent the total swarm size, and N_g and N_l denote the sizes of the *gbest* and *lbest* sub-swarms, respectively, with $N = N_g + N_l$. During the iterations, N_l is updated according to the following equation:

$$N_l = \frac{N}{2}(1 + e^{-3\sigma^2}) \tag{7.17}$$

where σ^2 represents the variance of the fitness values within the *gbest* sub-swarm, reflecting its convergence state. It is calculated using the following equation:

$$\sigma^2 = \sum_{i=1}^{N_g} (\frac{f_i - f_{avg}}{f})^2$$

$$f = \begin{cases} \max\{|f_i - f_{avg}|\}, \max\{|f_i - f_{avg}|\} > 1 \\ 1, \ others \end{cases} \tag{7.18}$$

where f_i represent the fitness value of the *i-th* particle in the *gbest* sub-swarm, and f_{avg} represents the current average fitness value of that sub-swarm. The variable f is a normalized scaling factor that helps limit the magnitude of σ^2. When σ^2 approaches 0, it indicates that the *gbest* sub-swarm is converging, N_l converges to $N/2$, resulting in the swarm's searching ability enhance. Conversely, when σ^2 tends toward infinity, N_l approaches to 0, resulting in a faster convergence speed for the swarm. Therefore, the adaptive adjusting the size of the *lbest* sub-swarm not only achieves a reasonable balance between global exploration and local convergence but also optimizes the contribution of each particle in the swarm, thereby improving the algorithm's operational efficiency.

In the absence of external interference, the *gbest* sub-swarm will rapidly converge as the number of iterations increases. To facilitate better collaboration between the two sub-swarms, the *gbest* sub-swarm must quickly incorporate the results from the *lbest* sub-swarm. During the iterative process, the *lbest* sub-swarm provides two types of information to the *gbest* sub-swarm: first, both sub-swarms synchronize their iterations to jointly update the historical *gbest* p_g of the swarm; second, if the *gbest* sub-swarm is converging (i.e., σ^2 is below the threshold σ_m^2), K best points in the neighborhood are selected from the *lbest* sub-swarm (where K represents the number of *lbest* sub-swarms, typically set as $K \approx \frac{N_l}{3}$ based on the referenced literature). This selection is used to update the historical optima of the K best particles in the *gbest* sub-swarm while retaining p_g and continuing the iteration.

Moreover, the *lbest* sub-swarm can also encounter premature convergence. Once the *lbest* sub-swarm converges prematurely, the entire swarm may converge rapidly with the iterations of the *gbest* sub-swarm. To mitigate this risk, if calculations show that $\sigma^2 < \sigma_m^2$ within the *lbest* sub-swarm, it will be reinitialized.

The steps of ATSPSO are as follows:

Step 1: Initialize all parameters and the initial positions of the particles;

Step 2: Calculate the fitness values for each particle in the swarm;

Step 3: Update the historical *pbest* p_i^k for each particle;

Step 4: Adjust the sizes of the two sub-swarms, N_g and N_l;

Step 5: If $\sigma^2 < \sigma_m^2$ in the *lbest* sub-swarm, reinitialize the *lbest* sub-swarm;

Step 6: Update the neighborhood best $p_{li}^j(i = 1, 2, ..., n; j = 1, 2, ..., K)p_{li}^j(i = 1, 2, ..., n; j = 1, 2, ..., K)$;

Step 7: If $\sigma^2 < \sigma_m^2$ in the *gbest* sub-swarm, select K neighborhood best from the *lbest* sub-swarm (where K is the number of *lbest* sub-swarms, typically set as $K \approx \frac{N_l}{3}$). Update the historical optima of the K best particles in the *gbest* sub-swarm, and retain p_g;

Step 8: Update the historical best value p_g^k of the swarm;

Step 9: Update the velocities and positions of the particles in both sub-swarms using the PSO iteration with linearly decreasing inertia weight.

Step 10: If the termination condition are met, stop the search; otherwise, return to Step 2.

In Pan [28], the algorithm was applied to solve nonlinear test functions, and the results demonstrated that ATSPSO has strong global search capabilities. It exhibited clear advantages, especially when optimizing complex problems with multiple peaks. However, during iterations, the algorithm requires calculating the variance of fitness values to assess the convergence of each sub-swarm, which, to some extent, impacts the overall efficiency of the algorithm.

7.4.5 Two Sub-Swarms PSO Based on Pheromone Diffusion

In real world, different species communicate in various ways, but the range of information diffusion between individuals is typically influenced by the strength of the signal and the distance between them. A classic example of information diffusion is the use of *pheromone* by both lower and higher animals for communication. Pheromones are chemicals produced by an individual's glands and released directly into the environment. These pheromones diffuse through mediums like air or water, allowing information to spread to other members of the population, transmitting signals such as mating cues, aggregation alerts, alarm signals, tracking, and marking information.

In Fan [29], to address the problem of PSO becoming easily trapped in *lbest* during the global optimization of multimodal, high-dimensional functions, the authors proposed a two sub-swarms PSO based on the mechanisms of pheromone diffusion and diversity feedback. After analyzing the reason of premature convergence, they divided the particle swarm into two co-evolving sub-swarms with opposing search directions. The introduction of a pheromone diffusion function controls the particles moving toward the swarm's current best position based on the relationship between the particles' positions, their corresponding fitness values, and the swarm's best position and fitness value. Additionally, the diversity feedback mechanism dynamically adjusts the inertia weight and regulates the number of particles in both sub-swarms.

Based on pheromone diffusion and diversity feedback mechanisms, the two sub-swarms PSO comprises three components: the incorporation of the pheromone diffusion function, dynamic adjustment of inertia weight, and co-evolution of the two sub-swarms.

① Pheromone diffusion function: In basic PSO, the global learning factor c_2 is applied uniformly, allowing all particles in a neighborhood to acquire the same "cognitive" ability from the current global best position $gbest$, regardless of their distance from it. This leads to rapid convergence toward $gbest$, thus reducing swarm diversity and increasing the likelihood of premature convergence. Fan [29] suggests that a particle's spatial position can be viewed as analogous to the distance between biological entities, while the improvement in its solution mimics the strength of pheromone signals. In order to address this, a regulatory coefficient is added to the learning factor c_2, creating a pheromone diffusion mechanism where particles closer to $gbest$ and showing greater solution improvement are more strongly attracted to $gbest$.

The learning factor c_2 is multiplied by the pheromone diffusion function H, defined as follows:

$$H_i = k \times (1 - \frac{d_i + \delta}{\max\limits_{1 \leq i \leq m} d_j + \delta}) \tag{7.19}$$

where k is defined as:

$$k = \begin{cases} \dfrac{s_i + \delta}{\max\limits_{1 \leq i \leq m} s_j + \delta} & if \ \dfrac{s_i + \delta}{\max\limits_{1 \leq i \leq m} s_j + \delta} > 0.6 \\ 0.6 & else \end{cases} \tag{7.20}$$

where H_i represents the degree of attraction that the current global best position ($gbest$) exerts on the i-th particle. H_i consists of two components. The first component, k, relates the particle's movement toward $gbest$ to the improvement in fitness from the particle's current value and $gbest$. Specifically, s_i represents the difference between the fitness value of the i-th particle and that of $gbest$. In addition, m represents the total number of particles, and s_j is the maximum difference in fitness values between any particle and $gbest$ within the swarm. Based on testing, k should not be set below 0.6. The second component links the movement of the particle to its Euclidean distance from $gbest$, where d_i is the Euclidean distance between the i-th particle and $gbest$, and d_j is the maximum Euclidean distance from any particle to *gbest* within the swarm. A small positive constant δ is added to prevent both the numerator and denominator from being zero simultaneously.

② Dynamic adjustment of inertia weight: Inertia weight is an important parameter in PSO for balancing global exploration and local exploitation. In the early stages of the algorithm, a larger inertia weight is required to promote broad exploration, while in the later stages, a smaller inertia weight is needed for fine-tuning around potential optimal solutions, with the added necessity of retaining the ability to escape *lbest*.

Apart from adjusting inertia weight based on the number of iterations, this method integrates swarm diversity into the adjustment process. This approach enhances local exploitation capabilities while reducing the risk of particles converging prematurely on *lbest*, which is especially critical for high-dimensional problems.

The diversity metric is defined as follows:

$$diversity(t) = \frac{1}{m \times |L|} \times \sum_{i=1}^{m} \sqrt{\sum_{d=1}^{D} (x_{id} - \bar{x}_d)^2} \tag{7.21}$$

The inertia weight is adjusted using the following equation:

$$\omega(t) = \omega_{final} + e^{-diversity(t)}(1 - \frac{Iter}{Iter_{max}}) \times (\omega_{initial} - \omega_{final}) \tag{7.22}$$

where *Iter* and *Iter*$_{max}$ represent the current number of iterations and the maximum number of iterations, while $\omega_{initial}$ and ω_{final} represent the initial and final inertia weights, respectively.

③ Co-evolution of two sub-swarms: In high-dimensional problems, the particle swarm is split into two sub-swarms, each evolving independently while sharing the same *gbest*. These two sub-swarms search for the optimal solution in opposite directions. This strategy broadens the search scope without increasing the number of particles, thereby reducing the likelihood of the algorithm getting trapped in *lbest*.

The update equation for basic PSO can be modified as follows:

$$V_i(t+1) = \omega(t) \times V_i(t) + c_1 \times r_1 \times (pBest_i(t) - X_i(t))$$
$$+ H_i \times c_2 \times r_2 \times (gBest(t) - X_i(t))$$
$$X_i(t+1) = X_i(t) + R \times V_i(t+1) \tag{7.23}$$

where R is defined as:

$$R = \begin{cases} 1, & if\,((diversity(t) < 0.3 \& P_i \le 0.3)\,or\,P_i \le 0.1) \\ -1, & else \end{cases} \tag{7.24}$$

When $R = 1$, particles search in the direction guided by both their *pbest* and the *gbest*. When $R = -1$, they search in the opposite direction. This adjustment allows for the independent evolution of particles in the primary and auxiliary sub-swarms. A random number P_i, generated between 0 and 1 for each particle, determines the probability of selecting a search direction. The algorithm ensures that at least 10% of particles belong to the auxiliary sub-swarm. If the swarm diversity drops below a threshold of 0.3, the number of particles in the auxiliary sub-swarm is increased to 30%.

The steps for the two sub-swarms PSO based on pheromone diffusion are as follows:

Step 1: Randomly initialize the positions and velocities of particles in the swarm, and divide the swarm into two sub-swarms: 90% as the primary sub-swarm and 10% as the auxiliary sub-swarm;

Step 2: Calculate the fitness value of each particle and set each particle's *pbest* and *gbest* of the swarm;

Step 3: Evaluate the diversity of the swarm and adjust the number of particles in both the primary and auxiliary sub-swarms.

Step 4: Update the velocities and positions of the particles, assigning the R value of the primary sub-swarm as 1 and the auxiliary sub-swarm as -1.

Step 5: Impose limits on velocity and position. If $V_i > V_{max}$ or $V_i < V_{min}$, set $V_i = V_{max}$ or $V_i = V_{min}$. Similarly, if $X_i > X_{max}$ or $X_i < X_{min}$, set $X_i = X_{max}$ or $X_i = X_{min}$;

Step 6: Calculate the fitness value of each particle at its new position;

Step 7: Update each particle's *pbest* position in both the primary and auxiliary sub-swarms;

Step 8: Update the *gbest* position in both the primary and auxiliary sub-swarms;

Step 9: Compare the fitness values of the two sub-swarms *gbest* positions, and select the position of the higher fitness value as the gbest position for the entire swarm;

Step 10: Check if the convergence criteria are met. If not, return to Step 3.

Step 11: Output the *gbest* particle's position and its corresponding optimal fitness value, and then terminate the algorithm.

The test results from the Fan [29] demonstrate that the two sub-swarms PSO algorithm based on the pheromone diffusion provides better stability. Compared to the standard PSO, this algorithm shows a stronger ability to escape *lbest*. The co-evolution of the two sub-swarms increases the likelihood of finding the optimal solution, leading to faster convergence. Throughout the process, swarm diversity remains within a controlled range, thereby reducing the risk of getting stuck in *lbest*.

7.5 Summary

This chapter provides an overview of various improved algorithms, such as PSO with topology-based improvements, PSO with improvements based on mathematical models, hybrid PSO, and PSO with multi-swarms. It also outlines the key improvement concepts and algorithm workflows for these improved methods.

References

1. Feng P, Xiaohui H, Eberhart R, et al (2008) An analysis of bare bones particle swarm. In: Proceedings of the 2008 IEEE swarm intelligence symposium. IEEE, Piscataway, NJ, USA
2. Kuo RJ, Syu YJ, Chen Z-Y, et al (2012) Integration of particle swarm optimization and genetic algorithm for dynamic clustering [J]. Inform Sci 195:124–140
3. Rui D, Xianbin F, Shuping L, et al (2012) Automatic generation of software test data based on hybrid particle swarm genetic algorithm. In: Proceedings of the 2012 IEEE symposium on electrical and electronics engineering (EEESYM 2012). IEEE, Piscataway, NJ, USA
4. Ali AM, Zendehboudi S, Lohi A, et al (2012) Reservoir permeability prediction by neural networks combined with hybrid genetic algorithm and particle swarm optimization [M]. https://doi.org/10.1111/j.1365-2478.2012.01080.x
5. Chen Y, Zhang D, Zhou M, et al (2011) Multi-satellite observation scheduling algorithm based on hybrid genetic particle swarm optimization. In: Proceedings of the 2nd International conference of electrical and electronics engineering, ICEEE 2011. Springer Verlag, Macau, China
6. Soleimani H, Kannan G (2015) A hybrid particle swarm optimization and genetic algorithm for closed-loop supply chain network design in large-scale networks[J]. Appl Math Model 39(14):3990–4012
7. Ghorbani N, Kasaeian A, Toopshekan A et al (2018) Optimizing a hybrid wind-PV-battery system using GA-PSO and MOPSO for reducing cost and increasing reliability[J]. Energy 154:581–591
8. Zhang Y, Wu L (2012) Restarted simulated annealing particle swarm optimization used in cluster analysis [J]. Int J Res Rev Soft Intell Comput 2(3):201–206
9. Bo S, Jiancang X, Ni W (2012) Application of urban water demand prediction model by using particle swarm algorithm based on simulated annealing [J]. Appl Mech Mater 155–156:102–106
10. Jiao B, Xu ZX (2012) Nonlinear inertia weigh particle swarm optimization combines simulated annealing algorithm and application in function and SVM optimization [J]. Appl Mech Mater 130(1):3467–3471
11. Tang M, Long C, Guan X et al (2012) Nonconvex dynamic spectrum allocation for cognitive radio networks via particle swarm optimization and simulated annealing [J]. Comput Netw 56(11):2690–2699
12. Niknam T, Narimani MR, Jabbari M (2012) Dynamic optimal power flow using hybrid particle swarm optimization and simulated annealing [M]. Europ Trans Electr Power https://doi.org/10.1002/etep.1633
13. Guo X, Sun HJ, Wu JJ et al (2017) Multiperiod-based timetable optimization for metro transit networks[J]. Transport Res Part B: Method 96:46–67
14. Xin B, Chen J, Zhang J, Fang H, Peng ZH (2012) Hybridizing differential evolution and particle swarm optimization to design powerful optimizers: a review and taxonomy [J]. IEEE Trans Syst Man Cybernet Part C Appl Rev 42(5):744–767
15. Niu SH, Ong SK, Nee AY (2011) A hybrid particle swarm and ant colony optimizer for multi-attribute partnership selection in virtual enterprises, pp 289–326. JohnWiley & Sons, Inc

16. Kiran MS, Ozceylan E, Gunduz M et al (2012) A novel hybrid approach based on particle swarm optimization and ant colony algorithm to forecast energy demand of Turkey [J]. Energy Convers Manage 53(1):75–83
17. Thanushkodi K, Deeba K (2012) Hybrid intelligent algorithm [improved particle swarm optimization (PSO) with ant colony optimization (ACO)] for multiprocessor job scheduling [J]. Sci Res Essays 7(20):1935–1953
18. Gholizadeh S, Fattahi F (2012) Serial integration of particle swarm and ant colony algorithms for structural optimization [J]. Asian J Civil Eng (Build Housing) 13(1):127–146
19. Wang LI, Zheng H (2012) A hybrid algorithm of PSO and GA for automatic placement of point annotation[J]. Comput Modern 10:30–37
20. Xie YP (2011) Research on improved particle swarm optimization algorithm and its application in control system design[D]. Beijing University of Chemical Technology
21. Tang XL (2007) Research on the theory and applications of chaotic particle swarm optimization algorithm [D]. Chongqing University
22. Sun H, Long T, Zhao J (2012) Swarm intelligence algorithm based on combination of shuffled frog leaping algorithm and particle swarm optimization[J]. J Comput Appl 32(2):428–431
23. Zhang Y, Liu YQ (2002) Soft computing methods [M]. Science Press, Beijing
24. Liu LJ (2008) Introduction of Tabu search in dual-swarm PSO and its application research [D]. Jiangnan University
25. Chang YW (2009) Research and application of vertical parameter multi-subswarm PSO [D]. China University of Mining and Technology
26. Cai W (1987) Matter element analysis [M]. Guangdong Higher Education Press, Guangzhou
27. Zhang HB (2010) Research on multi-swarm PSO based on extenics [D]. Beijing University of Chemical Technology
28. Pan ZM (2010) Adaptive twin swarms based particle swarm optimization [J]. Comput Appl Softw 7: 239–241+73
29. Fan HL, Zhong YC (2011) Two-subpopulation particle swarm optimization based on pheromone diffusion [J]. J Syst Simul 10:2125–2129

Chapter 8
Multi-objective PSO

In the mid-1980s, researchers began applying evolutionary computation to multi-objective optimization, and this resulted in the development of multi-objective evolutionary algorithms. Some of these algorithms have found successful applications in engineering practices, creating a new and rapidly growing area of research. Since multi-objective optimization problems do not possess a unique *gbest*, solving them involves identifying a set of optimal solutions. The application of PSO in multi-objective optimization is a significant research direction. This chapter primarily discusses the basic concepts of multi-objective optimization, benchmark functions, performance metrics, and the classification of multi-objective optimization methods, as well as the various categories of multi-objective PSO.

8.1 Multi-objective Optimization Problem (MOP)

Optimization techniques are mathematically based methodologies used to address various practical problems. In our daily lives, optimization problems are pervasive and play a crucial role. Multi-objective optimization is one of the primary research domains within optimization. Since the 1960s, the multi-objective optimization problem (MOP) has attracted increasing attention from researchers across diverse fields due to their widespread significance in real-world applications. By applying optimization techniques under consistent conditions, notable improvements can be achieved in system efficiency, energy consumption reduction, resource utilization, and economic benefits. For instance, the challenge of minimizing costs while maximizing benefits in factories is a classic example of MOP. Additionally, there are optimization and decision-making challenges associated with medium- and long-term development plans for social progress and national economies. In general, most

© Beijing Institute of Technology Press 2025
F. Pan et al., *Particle Swarm Optimizer and Multi-Objective Optimization*,
https://doi.org/10.1007/978-981-95-3381-7_8

optimization problems encountered in scientific and engineering practices are essentially MOPs and decision-making. These practical problems tend to be complex and difficult to resolve, necessitating considerable effort to find solutions. Therefore, addressing MPOs is a topic of significant scientific value and practical importance.

8.1.1 Development of MOPs

Most optimization problems, particularly during the design phase, necessitate the simultaneous optimization of multiple objective functions. For example, in unmanned aerial vehicle (UAV) mission assignments, the goal is to minimize combat duration while maximizing combat benefit and minimizing constraint violations, ideally achieving zero violations. In stock investments, we typically aim to minimize the amount of capital invested while maximizing returns and minimizing risks. Such problems involve multiple conflicting objectives, where a solution optimal for one objective may be sub-optimal for others, and this results in a set of trade-off solutions known as the *Pareto optimal set* or the *non-dominated set* [1]. The pursuit of the Pareto optimal set has been a central focus in both academia and engineering. Consequently, MOPs represent a widespread and complex class of optimization challenges encountered in optimization design. These problems often exhibit typical difficulties, including high dimensionality, discontinuity, non-convexity, polymorphism, or NP-hard. These features present significant challenges to existing deterministic methods, such as greedy algorithms, hill-climbing algorithms, branch-and-bound tree/graph search techniques, depth-first and breadth-first searches, best-first searches, and methods based on calculus.

Current literature on MOP methods published by various scholars can be divided into two categories: Methods for Single-objective-based MOPs (also known as traditional methods) and those for heuristic-based MOPs. The methods for Single-objective-based MOPs primarily combine multiple objectives into a single objective using established knowledge and find the resolution of that single objective to yield the optimal solution of MOPs. However, this results in a single-point solution. Common approaches include weighting methods, constraint methods, goal programming, and max–min methods. Nevertheless, classical multi-objective methods require decision-makers to possess prior knowledge of each objective before transforming multiple objectives into a single one. Furthermore, the solutions obtained are highly dependent on the weight vector or standard vector. These methods are also sensitive to the shape of the *Pareto* front. As the number of variables and constraints increases, the *Pareto* curve may exhibit non-convexity and discontinuity, significantly limiting the effectiveness of traditional methods.

With the development of various stochastic optimization techniques, methods for heuristic-based MOPs have emerged, including evolutionary algorithms, TS, and PSO. The swarm-based searching approaches enhance both multi-directional and global exploration, allowing for the handling of diverse objective functions and constraints. This has opened new avenues for solving MOPs, leading to the development of numerous multi-objective evolutionary algorithms. Significant research has focused on constructing Pareto sets, analyzing convergence, preserving distribution strategies, evaluating algorithm performance, and designing test functions.

Since *Schaffer* introduced the vector-evaluated genetic algorithm (VEGA) [2] in 1985, multi-objective evolutionary algorithms (MOEAs) have evolved through three distinct phases. The first phase, from 1985 to 1994, was characterized by slow development and included both non-*Pareto* and *Pareto* methods. Notable algorithms from this era include VEGA, multi-objective genetic algorithm (MOGA) [3], niched Pareto genetic algorithm (NPGA) [4], and non-dominated sorting genetic algorithm (NSGA) [5]. The second phase, spanning from 1994 to 2003, saw rapid advancements, marked by the integration of external archives and swarms into MOEAs. Key algorithms from this phase include the strength *Pareto* evolutionary algorithm (SPEA) [6], NSGA2 [7], Pareto Archived Evolution Strategy (PAES) [8], Pareto envelope-based selection algorithm (PESA) [9], and SPEA2 [10]. The third phase, from 2003 to the present, represents a comprehensive development period where new concepts, mechanisms, and strategies have been introduced into MOEAs to enhance performance and efficiency. These strategies include hybrid approaches, parallel methods, co-evolution strategies, dynamic evolution techniques, quantum evolution strategies, and differential evolution strategies. Furthermore, there has been preliminary progress in the study of high-dimensional MOPs and dynamic multi-objective optimization problems (DMOPs).

Compared to other evolutionary mechanisms, PSO offers significant advantages for solving MOPs. Firstly, PSO can effectively utilize clustering to search for non-dominated solutions in parallel, generating multiple non-dominated solutions during each iteration. Additionally, the memory function of PSO allows particles to track both the best solutions in the swarm and their own *pbest*, enhancing the search efficiency within the solution space. This leads to high computational efficiency and execution speed. Importantly, PSO is not dependent on the specific characteristics or structure of the problem, making it widely applicable to most multi-objective problems.

However, directly applying standard PSO to MOPs can result in particles tracking a non-Pareto optimal solution due to their inherent tracking characteristics. This can cause the algorithm to converge to a local range of non-dominated solutions without reaching the Pareto front. Therefore, when using PSO for MOPs, it is crucial to address the selection of both *gbest* and *pbest* positions. For selecting the *gbest*, the algorithm should achieve a good convergence speed while ensuring it converges to the Pareto front. In contrast, for selecting the *pbest*, we need to simplify the complexity introduced by this selection while optimizing the algorithm.

Currently, addressing MOPs is still in its early stages, with most efforts imitating techniques from other evolutionary algorithms. Existing approaches to MOPs can be broadly categorized into several types:

(1) Vector and weighting methods: Techniques like the adaptive weighting method and vector evaluation method proposed by Parsopoulos et al. [11] firstly applied

PSO to MOPs. However, selecting appropriate weight coefficients can be difficult, and vector evaluation often fails to yield satisfactory solutions.

(2) Pareto-based methods: Ray et al. [12] combined Pareto ranking with PSO to construct non-dominated solution sets. They used roulette wheel selection to choose the *gbest* particles. Lei et al. [13] further proposed the Pareto archive MOPSO by integrating archive maintenance with the *gbest* selection.

(3) Distance methods: Mostaghim et al. [14] allocated fitness values based on the distance between an individual's current solution and the Pareto solutions, selecting leading particles from the swarm. However, distance methods require initializing potential solutions, and if these initial values are too large, they can lead to insufficient selection pressure, resulting in slow convergence.

(4) Neighborhood methods: Hu et al. [15] introduced a selection strategy based on dynamic neighborhood, where one objective is defined as the optimization target and the others as neighborhood targets. This method can be sensitive to the choice of optimization target and the ranking of neighborhood objective functions.

(5) Multi-swarm methods: Pulido et al. [16] divided the swarm into multiple sub-swarms, each performing PSO independently. Information exchange between sub-swarms allows for the search of Pareto optimal solutions. However, this approach increases computational complexity due to the larger number of particles involved.

In summary, numerous methods have been developed to address MOPs, with researchers both domestically and internationally continuously proposing new strategies. Algorithms for MOPs have entered a phase of all-around development. However, research on multi-objective particle swarm optimization (MOPSO) is still at its starting stage, with substantial potential for future exploration, making it a valuable area for further study.

8.1.2   Mathematical Model and Basic Concepts of MOPs

MOPs, also known as multi-criteria optimization problems, typically involve optimizing several objectives simultaneously over n-dimensional decision variables and an m-dimensional objective space. An MOP can be expressed as follows [17, 18]:

$$min/max \, y = \boldsymbol{F}(\boldsymbol{x}) = (f_1(\boldsymbol{x}), f_2(\boldsymbol{x}), f_3(\boldsymbol{x}) \ldots f_m(\boldsymbol{x})) \tag{8.1}$$

$$s.t. \begin{cases} g_i(\boldsymbol{x}) \leq 0, i = 1, 2, 3, \ldots, q \\ h_j(\boldsymbol{x}) = 0, i = 1, 2, 3, \ldots, p \\ \boldsymbol{x} \in [\boldsymbol{x}_{min}, \boldsymbol{x}_{max}] \end{cases} \tag{8.2}$$

where X is the n-dimensional decision space, Y is the m-dimensional objective space, and $y = (y_1, y_2, \ldots, y_m)$ represents the m-dimensional objective vector. The objective function $F(x)$ defines m mapping functions from the decision space to the objective space, with $g_i \leq 0$ ($i = 1,2,...,q$) and $h_j(x) \leq 0$ ($j = 1,2,...,p$) representing the inequality and equality constraints of the problem, respectively. x_{min} and x_{max} represent the lower and upper bounds for the vector search.

Based on this mathematical model above, several key definitions have been developed in academic research [19].

Definition 8.1 Domination: A decision vector x_1 is said to dominate another decision vector x_2 if and only if x_1 is no worse than x_2 in all objectives, i.e., $f_k(x_1) \leq f_k(x_2)$ for all $k = 1,2,...,m$ (for minimization problems), and x_1 is strictly better than x_2 in at least one objective, meaning that there exists some $k = 1,2,...,m$ such that $f_k(x_1) < f_k(x_2)$. Similarly, an objective vector f_1 dominates another objective vector f_2, if f_1 is no worse than f_2 in all objectives and strictly better in at least one objective.

Definition 8.2 *Pareto* optimality: A decision vector x^* is *Pareto* optimal if there is no other decision vector $x \neq x^*$ that dominates it. In other words, for any $k, f_k(x^*) \leq f_k(x)$, If x^* is *Pareto* optimal, its corresponding objective vector $f^*(x)$ is also *Pareto* optimal.

Definition 8.3 *Pareto*-optimal set: The set of all *Pareto*-optimal decision vectors is known as the *Pareto* optimal set.

Definition 8.4 *Pareto* front: The surface formed by all *Pareto*-optimal objective vectors is called the *Pareto* front, as illustrated in Fig. 8.1.

Fig. 8.1 Definitions of MOPs in minimization

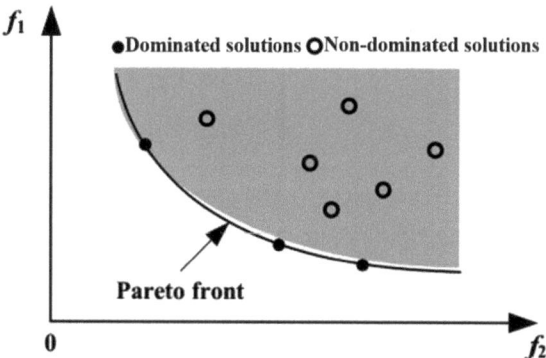

8.1.3 Benchmark Functions and Performance Metrics for MOPs

8.1.3.1 Benchmark Test Functions for MOPs

In order to effectively evaluate, compare, and classify algorithms and to enhance the algorithm performance for MOPs, thereby improving the efficiency and effectiveness of MOPSO, researchers have proposed a variety of benchmark test functions. These benchmark sets encapsulate key concepts in the field of multi-objective optimization [20], such as whether the MOP is continuous, discontinuous, or discrete; whether it is differentiable or non-differentiable; convex or concave; and whether the function is unimodal or multimodal.

The main benchmark test functions include ZDT1-ZDT6 and DTLZ1-DTLZ7, et al. The function expressions are as follows:

ZDT1: The *Pareto* front is convex.

$$\begin{cases} f_1(x) = x_1 \\ g(x) = 1 + 9(\sum_{i=2}^{n} x_i)/(n-1) \\ h(f_1, g) = 1 - \sqrt{\frac{f_1}{g}} \end{cases} \tag{8.3}$$

$x \in [0, 1]$. When $g(x) = 1$, the Pareto front is obtained.

ZDT2: The *Pareto* front is non-convex.

$$\begin{cases} f_1(x) = x_1 \\ g(x) = 1 + 9(\sum_{i=2}^{n} x_i)/(n-1) \\ h(f_1, g) = 1 - \frac{f_1}{g} \end{cases} . \tag{8.4}$$

$x \in [0, 1]$. When $g(x) = 1$, the *Pareto* front is obtained.

ZDT3: The *Pareto* front consists of multiple non-continuous convex surfaces.

$$\begin{cases} f_1(x) = x_1 \\ g(x) = 1 + 9(\sum_{i=2}^{n} x_i)/(n-1) \\ h(f_1, g) = 1 - \sqrt{\frac{f_1}{g}} - \frac{f_1}{g}\sin(10\pi f_1) \end{cases} \tag{8.5}$$

$x \in [0, 1]$. When $g(x) = 1$, $x_1 = [0,0.0830015349] \cup [0.1822287280,0.257762334] \cup [0.4093136748,0.4538821041] \cup [0.6183967944,0.6525117038] \cup [0.8233317983,0.8518328654]$, the *Pareto* front is obtained.

ZDT4: 219 local *Pareto*-optimal solutions are included, and they are used to test an algorithm's ability to handle multimodal functions.

$$\begin{cases} f_1(x) = x_1 \\ g(x) = 1 + 10(n-1) + \sum_{i=2}^{n}(x_i - 10\cos(4\pi x_i)) \\ h(f_1, g) = 1 - \sqrt{\frac{f_1}{g}} \end{cases} \quad (8.6)$$

$x_1 \in [0,1]$, $x_{i=2,\ldots,n} \in [-5, 5]$. When $g(x) = 1$, the *Pareto* optimal front is obtained.

ZDT6: The objective space is not uniformly distributed, therefore:

① Along the global *Pareto* front, the *Pareto* solutions are non-uniformly distributed;
② The closer a solution is to the *Pareto* front, the sparser the distribution; the further from it, the denser the distribution.

$$\begin{cases} f_1(x) = 1 - e^{-4x_1}\sin^6(6\pi x_1) \\ g(x) = 1 + 9\left((\sum_{i=2}^{n} x_i)/(n-1)\right)^{0.25} \\ h(f_1, g) = 1 - (\frac{f_1}{g})^2 \end{cases} \quad (8.7)$$

$x \in [0,1]$. When $g(x) = 1$ and $x_1 \in [0.2807753191,1]$, the *Pareto* front is obtained.

DTLZ1:

$$\begin{cases} \text{Minimize } f_1(x) = 0.5x_1x_2\ldots x_{m-1}(1 + g(x_m)) \\ \text{Minimize } f_2(x) = 0.5x_1x_2\ldots(1 - x_{m-1})(1 + g(x_m)) \\ \vdots \\ \text{Minimize } f_{m-1}(x) = 0.5x_1(1 - x_2)(1 + g(x_m)) \\ \text{Minimize } f_m(x) = 0.5(1 - x_1)(1 + g(x_m)) \\ \qquad g(x_m) = 100\left[|x_m| + \sum_{x_i \in x_m}(x_i - 0.5)^2 - \right. \\ \qquad \left. \sum_{x_i \in x_m}(x_i - 0.5)^2 - \cos(20\pi(x_i - 0.5))\right] \\ \qquad 0 \le x_i \le 1, i = 1, 2\ldots, n \end{cases} \quad (8.8)$$

DTLZ2:

$$
\begin{cases}
\textit{Minimize } f_1(x) = (1 + g(x_m)) \cdot \cos(x_1\pi/2) \cdots \cos(x_{m-2}\pi/2) \cdot \cos(x_{m-1}\pi/2) \\
\textit{Minimize } f_2(x) = (1 + g(x_m)) \cdot \cos(x_1\pi/2) \cdots \cos(x_{m-2}\pi/2) \cdot \sin(x_{m-1}\pi/2) \\
\textit{Minimize } f_3(x) = (1 + g(x_m)) \cdot \cos(x_1\pi/2) \cdots \sin(x_{m-1}\pi/2) \\
\vdots \\
\textit{Minimize } f_m(x) = (1 + g(x_m)) \cdot \sin(x_1\pi/2) \\
\qquad g(x_m) = \sum_{x_i \in x_m} (x_i - 0.5)^2, \, 0 \le x_i \le 1, i = 1, 2, ..., n
\end{cases}
\tag{8.9}
$$

DTLZ3:

$$
\begin{cases}
\textit{Minimize } f_1(x) = (1 + g(x_m)) \cdot \cos(x_1\pi/2) \cdots \cos(x_{m-2}\pi/2) \cdot \cos(x_{m-1}\pi/2) \\
\textit{Minimize } f_2(x) = (1 + g(x_m)) \cdot \cos(x_1\pi/2) \cdots \cos(x_{m-2}\pi/2) \cdot \sin(x_{m-1}\pi/2) \\
\textit{Minimize } f_3(x) = (1 + g(x_m)) \cdot \cos(x_1\pi/2) \cdots \sin(x_{m-1}\pi/2) \\
\vdots \\
\textit{Minimize } f_m(x) = (1 + g(x_m)) \cdot \sin(x_1\pi/2) \\
\qquad g(x_m) = 100 \left[|x_m| + \sum_{x_i \in x_m} (x_i - 0.5)^2 - \right. \\
\qquad\qquad \left. \sum_{x_i \in x_m} (x_i - 0.5)^2 - \cos(20\pi(x_i - 0.5)) \right] \\
\qquad 0 \le x_i \le 1, i = 1, 2..., n
\end{cases}
\tag{8.10}
$$

DTLZ7:

$$
\begin{cases}
\textit{Minimize } f_1(x) = x_1 \\
\textit{Minimize } f_2(x) = x_2 \\
\vdots \\
\textit{Minimize } f_{m-1}(x) = x_{m-1} \\
\textit{Minimize } f_m(x) = (1 + g(x_m)) \cdot h(f_1, f_2, ..., f_{m-1}, g) \\
\qquad g(x_m) = 1 + \dfrac{9}{|x_m|} \sum_{x_i \in x_m} x_i \\
\qquad h(f_1, f_2, ..., f_{m-1}, g) = m - \sum_{i=1}^{m-1} \dfrac{f_i}{1+g}(1 + \sin(3\pi f_i))
\end{cases}
\tag{8.11}
$$

8.1.3.2 Performance Metrics for MOPs

Performance metrics are used to evaluate the effectiveness of optimization algorithms. A variety of evaluation metrics have been proposed in academic research [21], including the generational distance (*GD*), the convergence metric (γ), and the diversity metric (Δ). These metrics can be roughly categorized into three types:

(1) Metrics that measure how closely the obtained solutions approximate the global *Pareto* front, primarily assessing the algorithm's convergence performance.
(2) Metrics that evaluate the diversity of the obtained non-dominated solutions.
(3) Composite metrics that assess both convergence and diversity, which will be discussed in detail in the next chapter.

8.1.4 Method Classification for MOPs

Since researchers began focusing on MOPs, a wide range of algorithms have been developed. Given the vast differences in methodologies and techniques, no unified classification standard currently exists. However, the two most common classifications are methods for single-objective-based MOPs and those for heuristic-based MOPs.

8.1.4.1 Methods for Single-Objective-Based MOPs

Methods for single-objective-based MOPs, also referred to as traditional optimization methods, involve transforming a MOP into a single-objective problem through certain predefined techniques. The resulting single-objective problem is then solved. Common approaches include the hierarchical sequence method, the constraint method, the effectiveness coefficient method (goal programming), and the evaluation function method.

(1) Hierarchical sequence method

The hierarchical sequence method ranks the objectives based on their importance. First, the solution to the highest-priority objective is obtained, followed by solving the next objective using the solution of the first, and this process continues sequentially. Instead of tackling multiple objectives simultaneously, this method solves them in order.

There are several variations of the hierarchical sequence method based on different hierarchical principles, such as the fully hierarchical method, the hierarchical evaluation method, and the key objective method. In the fully hierarchical method, each priority rank includes a single objective. While this method generally yields a satisfactory optimal solution, it has a limitation: if the optimal solution to an earlier objective is single, solving subsequent objectives becomes pointless.

(2) Constraint method

The constraint method involves selecting one of the sub-objectives from the multi-objective optimization problem as the new objective function while converting the remaining sub-objectives into constraints. For instance, if the selected sub-objective is the m-th objective, the other $(m-1)$ sub-objectives are transformed into constraint conditions.

The Mathematical formulation of the constraint method is as follows:

$$min/max\{f_1(x), f_2(x), f_3(x) \ldots f_m(x)\} \tag{8.12}$$

$f_1(x)$ satisfies:

$$f_2(x) < f_2^0(x), f_3(x) < f_3^0(x), \ldots, f_m(x) < f_m^0(x) \tag{8.13}$$

(3) Effectiveness coefficient method

The effectiveness coefficient method, also known as goal programming, is based on the principle that when optimizing objectives, the decision-maker first sets predetermined target levels for each objective function z_i ($i = l, \ldots, m$). Deviation variables are then calculated based on these target levels. Each objective function, in conjunction with its respective target level, forms an overall goal. In the case of minimization problems, the objective is expressed as $f_i(x) \leq z_i$, and the excess value σ_i above the target is minimized.

The Mathematical formulation of the effectiveness coefficient method is as follows:

$$min \sum_{i=1}^{m} w_i \sigma_i$$

$$s.t. \begin{cases} f_i(x) + \sigma_i \geq z_i, i = 1, \ldots, m \\ \sigma_i \geq 0, i = 1, \ldots, m \end{cases} \tag{8.14}$$

In weight-based goal programming, if the target levels represent *Pareto* reference points or if all deviation variables are positive, the resulting solution is guaranteed to be a *Pareto* solution. This method is straightforward and enables decision-makers to make informed choices easily. However, accurately setting the weight parameters can be challenging, and they may lack practical relevance.

(4) Evaluation function method

The evaluation function method is one of the most widely used approaches for solving multi-objective problems. Its fundamental principle is to utilize an evaluation function to aggregate and represent the relative importance of various objectives. The goal is to minimize this evaluation function to identify the optimal solution of the problem. Common techniques include the linear weighting method, the min–max method, and the ideal point method.

① Linear weighting method

The linear weighting method is one of the simplest and most commonly used approaches in multi-objective optimization. Its fundamental principle is to assign an inertia weight $w_i \geq 0$ ($\sum\limits_{i=1}^{m} w_i = 1$) to each objective function $f_i(x)$ based on its importance to the decision-maker. These weighted objective functions are then combined to create a new objective function $\sum\limits_{i=1}^{m} w_i f_i(x)$. By evaluating this new objective function, a solution to the MOP can be derived. Since the weights are constrained to $w_i \geq 0$, it can be demonstrated that the yielded solution is a non-dominated solution to the original MOP.

However, researchers have identified several significant drawbacks to this algorithm based on its principles. On the one hand, a small change in the weight parameters can lead to substantial variations in the objective vector. On the other hand, even if considerable changes in different weight parameters may produce similar solution vectors. Consequently, a uniformly distributed set of weights typically does not yield a uniformly distributed set of *Pareto* solutions.

② Min-max method

The min–max method operates on the principle that for a multi-objective minimization problem, the maximum value among the objective functions $f_i(x)$ is selected to construct the evaluation function, expressed as $z_i = max\{f_i(x)\}$. Therefore, the original problem can be reformulated as a numerical minimization problem: *min z = max(f_i(x))*.

Sometimes, to reflect the significance of each objective in the evaluation function, corresponding weights may also be assigned to the objectives, which can be mathematically described as follows:

$$\begin{cases} z_i = max\{w_i f_i(x)\} \\ min\ z = max(w_i f_i(x)) \end{cases} \tag{8.15}$$

③ Ideal point method

The ideal point method is predicated on the principle of identifying the location of the "ideal point" for the evaluation scheme. The optimal solution is then determined by minimizing the distance from each measured point to this "ideal point."

The mathematical formulation can be represented as follows:

$$min / max \{f_1(x), f_2(x), f_3(x) \ldots f_m(x)\}$$

$$f_j^* = min/max \left(f_j(x)\right), j = 1, 2, \ldots, m \tag{8.16}$$

Define the evaluation function as follows:

$$h(F(X)) = h(f_1, f_2, \ldots, f_n) = \sqrt{\sum_{j=1}^{m} (f_j(x) - f_j^*)} \qquad (8.17)$$

Then, we solve $min/max\ h(F(x))$.

The above methods for single-objective-based MOPs have achieved positive results in addressing practical problems of various fields. However, these methods have several significant drawbacks:

(1) In practical problems, the different physical meanings and measurement units of various objectives often prevent direct comparison or weighting. Although dimensionless normalization of objective functions can address this issue, it increases the complexity of the algorithm and may change the objective space, leading to ineffective use of decision information.
(2) These algorithms require users to provide precise decision information, but obtaining decision information that meets practical needs is often challenging, making it difficult to accurately establish the mathematical model for the optimization problem.
(3) Most algorithms can only yield the *lbest*. To avoid getting trapped in *lbest*, increasing the neighborhood size is a common approach; however, as the neighborhood expands, the complexity of the algorithm can increase exponentially.
(4) Many traditional methods are limited to relatively small problem sets, resulting in poor generality.

8.1.4.2 Methods for Heuristic-Based MOPs

Methods for heuristic-based MOPs are algorithms inspired by natural phenomena. These modern heuristic algorithms effectively avoid *lbest*, offer strong generalization capabilities, and exhibit high robustness. Unlike traditional optimization algorithms, which may be restricted to specific problem domains, modern heuristic methods, such as PSO and GA, use swarm-based search strategies. They perform parallel searches for potential solutions and can generate a *Pareto* solution set in a single run. Additionally, modern heuristics are not sensitive to the shape or continuity of the *Pareto* front, making them particularly well-suited for addressing real-world problems.

(1) Simulated annealing (SA)

SA is a well-known heuristic approach. Developed by *Kirkpatrick* et al. [22], it is a stochastic optimization algorithm that mimics the cooling process of metals in physics. SA exhibits strong capabilities in global search, making it highly effective for identifying non-dominated solution sets. The algorithm has been theoretically proven to converge to the *gbest* with a probability of 1, which is why it has been adapted for solving MOPs.

(2) Tabu Search (TS)

TS is another classic modern heuristic algorithm. Its key concept lies in maintaining a "tabu list" that records points previously visited. These points are then excluded from consideration in subsequent generations of solutions, thus ensuring diversified exploration and preventing the algorithm from getting stuck in *lbest*.

(3) Evolutionary algorithms (EAs)

EAs are swarm-based optimization methods. They work by evolving a swarm of potential solutions across generations, using probabilistic techniques to maintain diversity and conduct global searches. EAs do not require stringent mathematical conditions such as differentiability or convexity of the objective space, allowing them to handle a wide variety of objectives and constraints. This flexibility makes them particularly useful for solving complex real-world problems. As of September 2012, according to Mexican scholar *Coello* [23], there were over 7,200 English-language publications on the application of various intelligent optimization algorithms to solve MOPs. Commonly used EAs include multi-objective evolutionary algorithm (MOEA).

① Multi-objective evolutionary algorithm (MOEA).

MOEA is an extension of traditional EAs, designed specifically to solve MOPs. Unlike single-objective-based EAs, MOEAs aim to obtain a set of non-dominated solutions that not only approximate the global Pareto front but are also evenly distributed across it. Different MOEAs employ various methods to achieve this, making it difficult to describe them within a single framework. For clarity, a general process of a Pareto-based MOEA is provided below:

General processes of a *Pareto*-based MOEA:
Step 1: Initialize swarm P;
Step 2: Evolve swarm P using EA to generate offspring R;
Step 3: Extract the non-dominated solutions from $P \cup Q$ and store them in an external archive Q;
Step 4: Adjust the size of the external archive Q and maintain its diversity;
Step 5: Check if the termination conditions are met. If not, proceed to Step 6. Otherwise, output archive Q as the result;
Step 6: Set $P = Q$ and return to Step 2.
A flowchart of the process is shown in Fig. 8.2.

For MOPs, it is difficult to define an optimal stopping criterion. Researchers typically set a predefined number of iterations or function evaluations as the termination condition.

Several important considerations must be addressed in MOEA:

(1) Preserve the non-dominated solution set from the previous generation to participate in the next generation. This is similar to retaining the *pbest* in traditional EAs, ensuring that the non-dominated set in the new generation is at least as good as the previous one, which is a necessary condition for convergence;

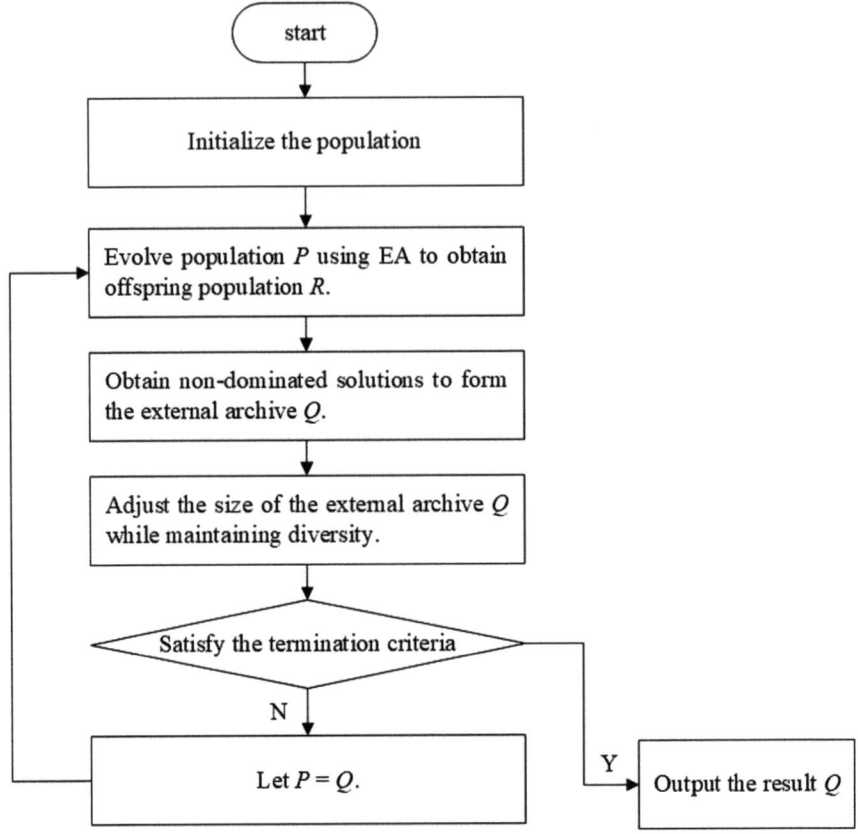

Fig. 8.2 A flowchart of MOEA

(2) Determining how to construct the non-dominated solution set;
(3) Devising a strategy to control the size of the non-dominated solution set;
(4) Maintaining diversity within the non-dominated solution set.

② Multi-objective particle swarm optimization (MOPSO)

MOPSO is an adaptation of PSO to solve multi-objective problems. Similar to
EAs, MOPSO uses swarm-based search methods and information exchange during
the search process, resulting in several similarities between MOPSO and MOEAs.
However, there are key differences: in MOPSO, there are multiple *gbest* positions
and *pbest* positions for each particle during iterations in the swarm, which must
be selected using appropriate strategies. Moreover, MOPSO employs an external
archive to store *gbest* and *pbest*, which is a feature absent in most MOEAs. This
requires the archive's size and diversity to be carefully considered.

  General processes of a *Pareto*-based MOPSO:

Step 1: Initialize particle positions x and velocities v, calculate each particle's objective vector, and add the non-dominated solutions to the external archive NP;

Step 2: Determine the initial *gbest* and *pbest*;

Step 3: Update particle velocities v and positions x, while ensuring they remain within the search space, compute each particle's objective vector, and adjust *pbest* accordingly;

Step 4: Maintain the external archive NP based on the new non-dominated solutions and select a new *gbest* for the swarm;

Step 5: Check if the termination conditions are met. If not, return to Step 3. Otherwise, output the final results.

A flowchart of the process is shown in Fig. 8.3.

Although MOPSO is not as mature as MOEAs, and remains in the early stages of research, its simplicity, ease of implementation, and deep foundation in intelligent

Fig. 8.3 A flowchart of MOPSO

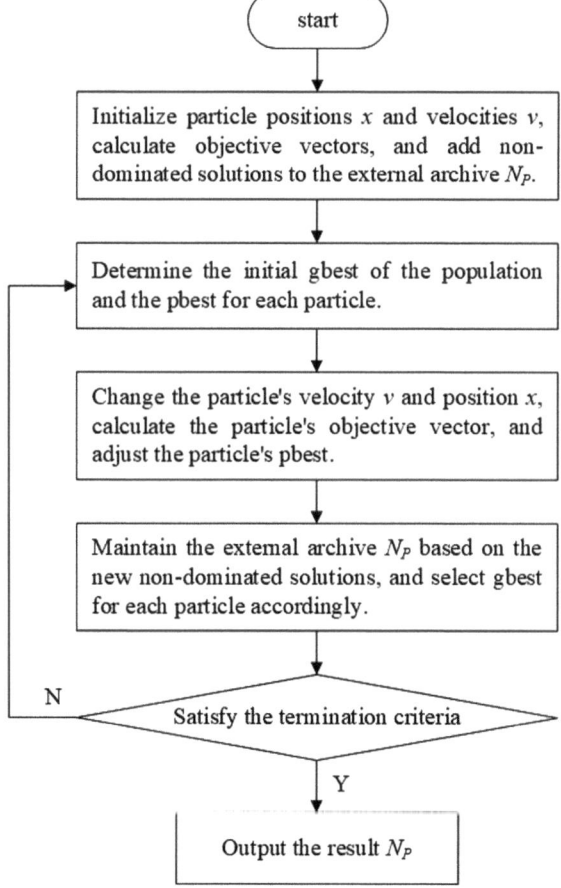

systems make it highly suitable for both scientific research and engineering applications. Exploring new applications for PSO is a valuable endeavor, as its true strength lies in its potential for engineering use.

8.2 Classification of MOPSO

8.2.1 Based on Different Selection Mechanisms

In the single-objective optimization problems, the swarm topology is fixed, so the solutions to *gbest* and *pbest* are uniquely determined. However, in MOPs, the non-dominated solution space contains multiple potentially optimal particles, leading to different swarm topologies and various strategies for selecting the *pbest* in the swarm. Therefore, determining how to select the optimal particle from all known non-dominated solutions becomes a key research focus.

(1) Random selection

When dominance relationships cannot be used to select a solution from the non-dominated set, random selection is the simplest and most direct approach, which was commonly used in the early stages of research. One of the easiest methods for selecting *gbest* from the external archive is to randomly choose one non-dominated solution [24–26], with each solution having an equal selection probability. However, this can result in higher selection probabilities in regions where particles are overly concentrated, thus leading to poor distribution along the *Pareto* front and reduced swarm diversity. In Moore et al. [25], Moore and Chapman [27], all currently identified non-dominated solutions are stored in *p-lists*. The *pbest* for each individual is randomly selected from these *p-lists*, and by comparing it with the non-dominated solutions in the *p-lists*, the best neighboring particle (*nbest*) is identified, which is also a non-dominated solution.

(2) *Pareto* dominance-based *pbest* selection

Compared to the wide variety of strategies for selecting *gbest* (or *lbest, nbest*), many papers propose similar strategies for selecting *pbest*, with shared characteristics. *Pareto* dominance-based *pbest* selection is a relatively simple and commonly used approach. For example, in Li [24], Moore et al. [25], Raquel and Naval [25], Hsieh and Al [26], Koduru et al. 28], the non-dominated particle is chosen as *pbest* based on a comparison between the current particle position and its historical best position. If neither position dominates the other, one is selected randomly. In Raquel and Naval [26], Hsieh and Al [28], Coello et al. [30], Hu et al. [31], *pbest* is updated only if the new solution dominates the current *pbest*; otherwise, no update or selection is made. In Koduru et al. [33], the method differs slightly in that if the solutions cannot be compared or are mutually non-dominant, the current position is chosen as the *pbest*.

(3) *Sigma* selection strategy

Mostaghim et al. [32] proposed the *Sigma* method to determine the *gbest* position for each particle. In this method, if a particle's *Sigma* value is closest to the *Sigma* value of a member in the external archive, that member is selected as the particle's *gbest* position.

(4) Dynamic neighborhood selection.

Hu et al. [30] proposed a dynamic neighborhood strategy to determine the *gbest* position for particles. In their approach to bi-objective optimization, the selection process for a particle's *gbest* is as follows: First, the distance between the particle and other particles is calculated based on the first objective. From these distance values, k locally optimal solutions are identified. Finally, the solution with the best performance in the second objective is chosen as the particle's *gbest* from the k solutions.

(5) Roulette wheel selection.

Coello Coello et al. [34] introduced a fitness value for each grid that contains at least one individual from the external swarm. The roulette wheel method is then used to select a grid, and from this grid, one particle is randomly chosen as the *gbest* position.

Salazar—Lechuga et al. [35] incorporated the concept of fitness sharing into MOPSO. Each solution in the external archive is assigned a shared fitness value. The roulette wheel method is then employed to select the *gbest* for each particle based on the size of the shared values.

(6) Probabilistic selection of optimal particles.

Ray [12] proposed a probabilistic method for selecting the optimal particle, with the probability largely determined by the crowding radius. This method promotes the exploration of new regions and enhances the distribution along the *Pareto* front. The crowding radius refers to the average distance between a particle and its nearest neighbors in the objective space. The fewer particles surrounding an optimal particle, the higher the probability of its selection. In essence, this measure, similar to the metric proposed by *Carlo* and *Ray*, represents the average distance between adjacent solutions.

(7) Niche comparison selection.

Liu [36] applied a niche comparison method to select the *gbest* particle from the archive by favoring those with smaller niche counts. In order to enhance diversity, particles in densely populated regions (i.e., areas with high niche counts or short crowding radius) are eliminated, and a new particle is randomly generated in their place.

Table 8.1 Classification of PSO in MOPs

	Category	Characteristics	Methods
(1)	Priori methods	Decision → Optimization search	Weighted aggregation method, lexicographic method, min–max method
(2)	Posteriori methods	Optimization search → Decision	General methods; distance-based methods; grid-based methods; clustering-based methods; hierarchical ranking methods; sub-swarm methods
(3)	Extended methods	Fitting the fitness landscape	Fitness inheritance; fitness estimation
(4)	Performance metrics	Nine different metrics in the *Pareto* sense	

8.2.2   Based on Different Decision Mechanisms

The MOPSO algorithm can be classified into four main categories, as shown in Table 8.1.

8.2.2.1   Priori Methods

Priori methods specifically refer to optimization techniques that use prior knowledge or information about multiple objectives before conducting the optimization search [37]. Based on the preferences of the decision-maker, the sizes of the weight vectors for the multiple objectives are established, which means that decisions are made before optimizing the MOPs. For the optimization of MOPs, enhancements to PSO primarily incorporate three types of a priori methods: the weighted aggregation method, the lexicographic method, and the min–max method.

(1)   **Weighted aggregation method**

In the weighted aggregation method, all objective functions are combined into a single-objective problem through a weighted aggregation, where the weights reflect the contribution of each objective function to the overall goals, followed by normalization. This method is considered one of the simplest ways to address multi-objective problems. The expression for the aggregated fitness function $f_i(x)$ is as follows:

$$F = \sum_{i=1}^{k} w_i f_i(x) \tag{8.18}$$

where w_i represents the weight and satisfies $\sum_{i=1}^{k} w_i = 1$. During the optimization process, the values of w_i can be initially set or dynamically adjusted based on specific strategies. Parsopoulos [38] employed three fixed or adaptive weighting methods to convert MOPs into single-objective ones and provided a comprehensive analysis and comparison.

① **Conventional Weighted Aggregation**

Conventional weighted aggregation (CWA) uses fixed weights and requires suitable weights to be selected based on prior knowledge. Xia [39]employed a linear weighting method to solve a three-objective scheduling problem and ultimately obtained feasible solutions for each objective. *Zhang* [40] applied a fixed-weight PSO to design a PID controller. Since each trial can yield only one Pareto-optimal solution, multiple optimization runs are necessary to search for a sufficient number of Pareto-optimal points. However, in most cases, it remains uncertain how many Pareto-optimal points exist, significantly limiting the effectiveness of this method. Additionally, due to time constraints and the substantial computational workload from repeated calculations, few studies employ the CWA for solving MOPs.

② *Bang-Bang* **Weighted Aggregation**

The *Bang-Bang* weighted aggregation (BWA) is an algorithm that adjusts weights during optimization through a weight update formula. The equation for updating weights is as follows:

$$w_1(t) = sign(sin(2\pi t/F)), w_2(t) = 1 - w_1(t) \tag{8.19}$$

We can see that the weight adjustments are abrupt. While BWA serves as a prototype improvement over the dynamic weighted aggregation (DWA) and CWA, few studies have combined it with PSO to solve MOPs, apart from the algorithm used for experimental comparisons in Konstantinos and Michael [41].

③ **Dynamic Weighted Aggregation**

The dynamic weighted aggregation (DWA) adjusts its weight values throughout the optimization process to continuously approach the Pareto front. Liu [42] used the modified fuzzy-Chebyshev Programming (MFCP) to generate weights, quantifying the importance of each objective based on its satisfaction level. Marandi [43] applied a weighted mean ranking method to transform a three-objective problem into a single-objective function, deriving weight sizes using a Mamdani fuzzy inference system. Baumgartner [44] employed a gradient method to adjust weight sizes, effectively dividing the swarm into multiple sub-swarms. Each sub-swarm refers to its optimal particle for searching, and the overall result for the swarm is obtained through weighted aggregation.

The interactive weighting method is a type of dynamic weighting technique where the values of the weight vectors are dynamically adjusted through generation or adjustment rules during the algorithm's execution, allowing for some degree of interaction between the preference of decision-maker and the algorithm. A typical example of this approach is the decomposition method.

The decomposition method involves transforming the MOPs into multiple single-objective optimization problems solved simultaneously with different weight vectors. For an MOP with m objectives, if the weight vector is $\Lambda = (\lambda_1, \lambda_2, \cdots, \lambda_m)$, the Chebyshev decomposition strategy can be used to convert the MOP into the following single-objective problem:

$$\min\{\min\{g(x|\Lambda, z)\}\} = \max\{\max\{\lambda_j |f_j(x) - z_j|\}\}, 1 \leq j \leq m \qquad (8.20)$$

where $\lambda_j \geq 0 (j = 1, 2, \cdots, m)$, $\sum_{j=1}^{m} \lambda_j = i$, and $z = (z_1, z_2, \cdots, z_m)$ represents the reference points for optimization. In minimizing MOPs, different weight vectors are typically assigned to each particle, with nearby particles selected based on similar weight settings to form neighborhoods.

In 2007, Zhang and Li [45] examined three methods: linear weighting, weighted *Tchebycheff*, and boundary crossover weighting. They employed a scalarization approach that decomposes MOPs into multiple single-objective-based optimization problems using various weights. Additionally, they introduced a strategy for optimization search based on individual information within the weighted neighborhood, known as the MOEA/D algorithm. This algorithm is capable of generating multiple *Pareto*-optimal solutions in a single run and achieves remarkable results. Li and Zhang [46] further proposed the MOEA/D-DE algorithm, which utilizes the *Tchebycheff* decomposition strategy to transform MOPs into several single-objective-based problems. Subsequently, based on the individual information in the weighted neighborhood, the differential evolution (DE) method is employed to evolve all individuals. Similarly, Peng and Zhang [47] applied PSO to simultaneously search for the scalar optimal values of multiple decomposed single-objective-based optimization problems. Martínez and Coello Coello [48] proposed a decomposition strategy characterized by penalty boundary crossover and employed PSO to solve the decomposed single-objective-based optimization problems.

All types of weighting methods can be derived using the *Kuhn-Tucker* conditions. Furthermore, weighting methods require that the *Pareto* front be a convex set, which is essential for obtaining non-dominated solutions. Simply assigning weights to each objective and obtaining a single optimal solution is generally insufficient for solving MOPs. There are at least three major issues: first, how to retain previously identified solutions while continuing to explore the unknown; second, whether there is a need for dynamic weights that characterize the relationships between objectives and how to design dynamic strategies that guide particles toward the *Pareto* front; and finally, as the number of objective functions increases, the number of weights and the size of the solution space will grow significantly, which raises concerns about the computational feasibility of using dynamic weights to transform MOPs into single-objective ones. In fact, the weighted aggregation method is not an effective or direct approach to addressing the fundamental issues of MOPs.

(2) Lexicographic method

In the lexicographic method, objectives are ranked based on certain factors (such as importance, complexity, priority, etc.), and optimization is performed sequentially following the order of the ranked objective functions.

Hu [30] introduced the dynamic neighborhood particle swarm optimization (DNPSO) and integrated the lexicographic method into MOPSO. DNPSO first designates one objective as the primary optimization target, with the remaining objectives treated as secondary or neighboring targets. Optimization is first applied to the first objective function (which might be the most important or highest priority). The particles in the nearest neighborhood are grouped according to their distance from the fitness value of the first objective function, which can also be considered a clustering process. Following the optimization of the first objective, the second objective is then optimized, and this process continues iteratively.

In addition, Hu [31] employed extended storage to preserve non-dominated solutions discovered during optimization, which followed the same selection mechanism. Combining extended storage with the dynamic neighborhood strategy improved the uniform distribution of solutions along the *Pareto* front, thus enhancing swarm diversity.

The lexicographic method is computationally efficient with a low complexity when the number of objectives is small. However, selecting and prioritizing objectives requires prior knowledge of the objective functions, as the order of the objectives significantly impacts the results. Furthermore, this method assumes that the objectives are orthogonal. As the number of objectives N (>3) increases, the computational burden rises exponentially $\left(O\left(N^{N-1}\right)\right)$.

(3) Min–max method

Li introduced the the maxmin strategy from game theory into MOPSO, proposing an improved maxmin PSO algorithm. The fitness function of maximin PSO is derived from the maxmin strategy, and it relies on *Pareto* dominance and the ranking of individuals within the swarm without requiring clustering or niche methods to replace the ranking of dominated individuals.

8.2.2.2 Posteriori Methods

Posteriori methods refer to optimization techniques that do not require prior knowledge or preprocessing of the problem. In multi-objective optimization, posteriori PSO algorithms are primarily based on *Pareto* optimality. These methods generate a set of *Pareto*-optimal solutions after the optimization, which are presented to decision-makers for selection. The enhancement and application of PSO in multi-objective problems through posteriori methods focus on three key areas: non-dominated solution sets, optimal particles, and diversity (Fig. 8.4). Each of these areas is further divided into more detailed research topics, represented by the six outer segments of

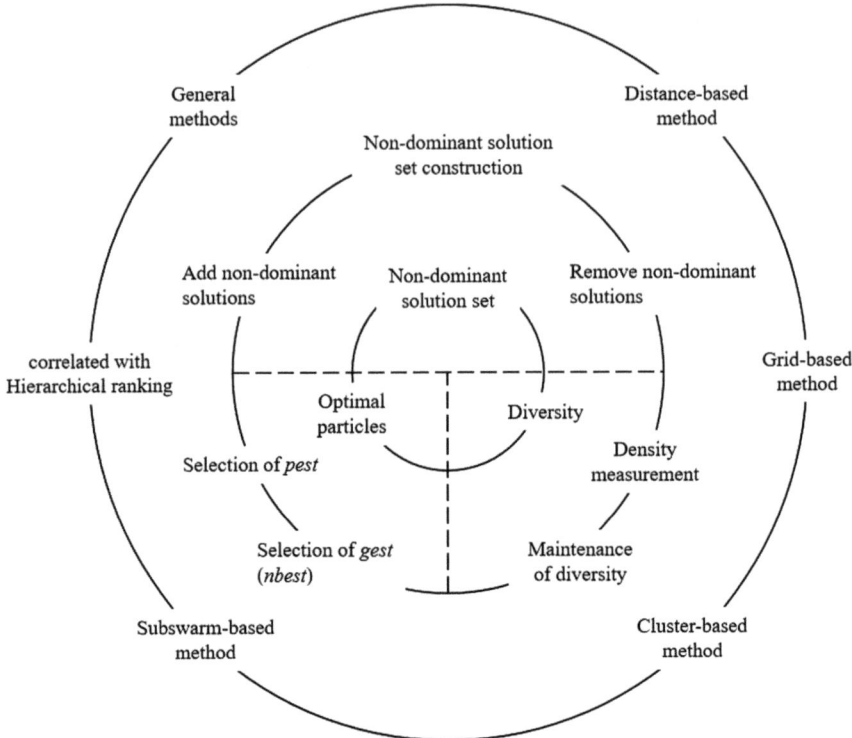

Fig. 8.4 Classification structure of posteriori methods

the diagram. Research on non-dominated solution sets, optimal particles, and diversity typically overlaps with these six subtopics. This means that some studies may reappear in multiple contexts throughout the discussion.

Common posteriori methods include general methods, distance-based methods, grid-based methods, clustering-based methods, rank-based methods, and sub-swarm-based methods. The specific descriptions are as follows:

(1)   General methods.

① **Non-dominated solutions.** In single-objective-based optimization, a particle's *pbest* refers to the best position it has encountered so far, while *gbest* is the best position found by the entire swarm during the search. These optimal solutions are represented as points or vectors in the solution space, and their fitness values can be directly compared. However, in MOPs, the optimal solution is no longer a single point but rather a *Pareto*-optimal solution set. In this case, *gbest* is selected from a set of non-dominated particles, and *pbest* is no longer judged by optimality but based on dominance relations. For non-dominated solutions, they are all equally considered optimal. In PSO algorithms for MOPs, it is essential to store all non-dominated solutions discovered during the search. Depending on how they are stored, non-dominated solution sets can be classified into internal and external archives. Additionally, based on their capacity design, they can be divided into fixed-size archives and unbounded archives.

For internal archives, non-dominated solutions within the swarm are saved without requiring any additional data structures. For example, a currently discovered non-dominated solution is stored at its *pbest*. In PSO algorithms commonly applied to MOPs, the population size is typically fixed. Consequently, the internal archive has a strictly defined storage capacity, with the number of non-dominated solutions in the swarm equaling the swarm size. This implies that opting for a larger swarm size is advantageous for preserving a greater number of *Pareto* solutions. In early research adapting PSO to solve MOPs, Moore proposed using *pbest* to track all *Pareto* non-dominated solutions during the particles' exploration of the search space. This modification was crucial in making PSO suitable for multi-objective optimization. Hu [30] further developed this concept by storing all potential *pbest* in a sufficiently large archive. This approach enables the algorithm to identify non-dominated solutions by examining the set of *pbest* within the archive.

The external archive refers to storing historical best solutions as a set of non-dominated solutions in additional memory space allocated by the algorithm rather than within the algorithm's population. These historical best solutions are often referred to as the repository, external archive, or extended memory. The size of this storage space can be either fixed or adaptively adjusted. Only a few papers use dynamically adaptive data dimensions, while most external archives have a strictly defined storage capacity. In a fixed-size archive, when the archive reaches its capacity limit, some non-dominated solutions must be discarded. Common criteria for discarding solutions include density distribution, selection pressure, and clustering levels. Therefore, for MOPs, a critical issue for PSO is how to construct an effective storage structure for non-dominated solutions and how to maintain this set (adding or removing solutions).

② **Optimal particles**. In single-objective constrained optimization problems, the *gbest* and *pbest* are uniquely determined due to the defined swarm topology. In MOPs, the non-dominated solution space contains numerous potential optimal particles, and this leads to various possible swarm topologies and diverse strategies for selecting the best particles within the swarm. In addition, numerous selection strategies and computational methods, such as round-robin comparison, roulette wheel selection, and niching, have been discussed in existing literature. Thus, preserving the non-dominated solutions found by the particles and selecting the optimal individuals from all known non-dominated solutions are key areas of research closely related to optimal particles.

When confronted with a set of non-dominated solutions that cannot be differentiated based on dominance relationships, random selection emerges as the simplest and most direct method. This approach was commonly employed in early studies. The most straightforward method for selecting the *gbest* from an external archive is to randomly choose one non-dominated solution [24, 25, 49]. In this approach, each non-dominated solution has an equal probability of being selected. However, this approach may result in higher selection probabilities in regions where particles are overly concentrated, which can hinder the distribution of particles along the *Pareto* front and reduce swarm diversity. In Moore et al. [25], Moore and Chapman [27], all currently discovered non-dominated solutions are stored in individual particle lists (*p-lists*), and the *pbest* is randomly selected from these lists. By comparing with the non-dominated solutions in the *p-lists*, the neighborhood best (*nbest*) is determined, which is also a non-dominated solution. However, most papers propose various methods for selecting the *gbest* or *nbest* particles, which will be summarized and discussed in the following sections.

In comparison to the various strategies for selecting *gbest* (*lbest*, *nbest*), many papers adopt similar approaches for choosing the *pbest*. These strategies often share common characteristics. A relatively simple and commonly used method for *pbest* selection is based on *Pareto* dominance. For example, in Li [24], Moore et al. [25], Carlo and Navaljr [49], non-dominated particles are chosen as the *pbest* from either the particle's current position or its historical best. If neither position dominates the other, one is randomly selected. In Hu and Eberhart [15, Li [24], Carlo and Navaljr [49], *pbest* is only updated if the new solution dominates the current *pbest*; otherwise, no update occurs. In Moore et al. [25], there is a slight variation: if the solutions are non-comparable or mutually non-dominating, the current particle position is selected as the *pbest*.

③ **Density measurement and diversity preservation**. Unlike single-objective optimization which seeks a unique optimal solution, the main goal in multi-objective optimization is to find a *Pareto*-optimal set of solutions. Throughout the optimization process, all particle updates are guided by the non-dominated solutions already discovered and stored in the archive. In early studies applying PSO to MOPs [35, 50], the issue of diversity preservation was not addressed. However, the distribution of solutions in the objective space is crucial. Ideally, particles should be distributed as widely, numerously, and uniformly as possible along the *Pareto* front. Therefore, maintaining swarm diversity becomes a critical challenge. One approach is to manage the selection of non-dominated solutions to prevent particles from clustering excessively around the best particles, a topic that will be categorized and discussed later in this section. Another strategy involves enhancing PSO's update equations and selection mechanisms by integrating operations from EA, GA, or other optimization techniques (Table 8.2).

Table 8.2 Improvements in update equations and selection strategies	Improvement methods	References
	Roulette wheel	[21, 24, 42–48]
	Mutation	[21, 24, 25, 27, 35, 49–51]
	Crossover	[24, 51]
	Turbulence	[21, 24, 28, 49, 52, 53]

Ho [54] proposed a typical improvement to the algorithm by introducing operations with random characteristics to enhance the update equations. *Ho* considered individual emotion, social experiences, velocity reversal, and directional mutations. A "craziness variable" was introduced, which acts as a disturbance operation with a random magnitude aimed at maintaining the diversity of the particles in the optimization algorithm. In addition to these four random parameters, an age variable for the solutions stored in the archive was introduced to further enhance the algorithm's diversity. The selection of solutions is based on their age variables and adaptive weights, employing a roulette wheel selection method.

While the roulette wheel selection mechanism is designed to give equal selection probability to all *Pareto* individuals, in practice, only a few densely populated regions of non-dominated solutions receive higher probabilities, which is detrimental to maintaining swarm diversity. In order to address this issue, a method combining roulette wheel selection with particle distribution density in the archive has been developed to mitigate this selection bias. The unidirectional information transfer in PSO, along with the broadcast-style information-sharing structure under the classic PSO social model, distinguishes PSO from GA through its unique information interaction approach. This method allows PSO to converge quickly but poses challenges for maintaining diversity. Consequently, various social (information) topologies have been incorporated into PSO's improvements, along with operations commonly used in EA and GA, such as different forms of mutation and crossover operations, aimed at altering the patterns of information transfer. Disturbance operations, similar to mutation in evolutionary strategies, involve adding Gaussian white noise to randomly selected particle position vectors (or velocity vectors) in the update equations, helping particles escape *lbest* and explore other unknown regions.

(2) **Distance-based methods**

Carlo [26] introduced a crowding distance mechanism among particles into the MOPSO and proposed the crowding distance with multi-objective PSO (MOPSO-CD). A fixed external archive is designated to store the discovered non-dominated solutions. All newly explored non-dominated solutions that are not dominated by any already stored solutions are saved in a variable, while any dominated solutions within that variable are removed. If the maximum dimension of the variable is reached, the most densely packed particles will be eliminated. *Carlo* uses the crowding distance of the solutions to evaluate their density, which describes the size of the largest hypercube surrounding a particle, excluding any other particles (as illustrated in Fig. 8.5 for particle *i*). All non-dominated solutions in the variable are arranged according to their crowding distances, and one particle is randomly selected as the *gbest* for each particle from a specified portion (e.g., the top 10%). A particle is also randomly selected from the bottom specified range of densely populated particles and removed from the archive.

Ray [12] proposed a probabilistic method for selecting optimal particles, where the selection probability primarily depends on the aggregation radius, thus allowing exploration of new areas to improve distribution along the Pareto optimal boundary. The aggregation radius refers to the average distance between a particle and its

Fig. 8.5 Crowding distance
of adjacent solutions

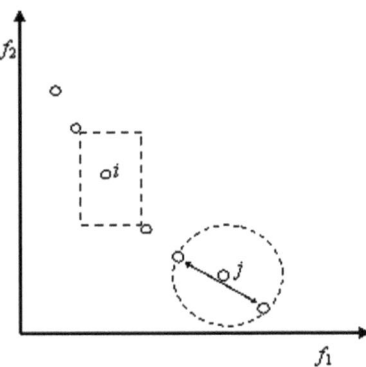

neighboring particles in the objective space (as shown for particle j in Fig. 8.5).
The fewer particles surrounding the optimal particle, the higher the probability of
its selection. In fact, similar to *Carlo*'s metric, *Ray*'s metric represents the average
distance between two adjacent solutions.

Li [55] introduced a non-dominated sorting PSO method incorporating two free-
parameter niche techniques. This approach compares the *pbest* positions of all particles with their offspring, followed by the classification and sorting of the entire
swarm. This method ensures the discovery of a more comprehensive set of non-
dominated solutions while guiding the swarm toward the *Pareto*-optimal set through
appropriate selection pressure. Two niche methods are employed: Niche count and
crowding distance assignments. The niche count is determined by the number of
other particles within a specified radius (Fig. 8.6). Conversely, the crowding distance
of individual particles represents the average distance to surrounding particles in
each objective space, aligning with the method used by *Ray* and *Carlo* (Fig. 8.5).
Particles in less dense regions have smaller niche counts, with the particle possessing
the smallest niche count selected as *gbest*. In the crowding distance method, the algorithm calculates the crowding distance for each solution particle, which is defined as
the average distance between two adjacent particles along each objective function. All
particles in the current *Pareto*-optimal set are then sorted in descending order based
on their crowding distances. The *gbest* for each particle is randomly selected from
the top portion of this sorted list, favoring particles with higher crowding distances.
Analogous to this sorting and random selection approach, Liu [36] employed a niche
comparison to select the *gbest* particles with small niche counts from the archive.
In order to enhance diversity, particles in high-density areas (characterized by large
niche counts or small crowding distances) are removed and replaced with randomly
generated new particles.

Zhang [56] employed two storage variables similar to those in Abido [51] to
record *pbest* and *gbest*, with operations on these variables occurring in two distinct
phases. In the early stages of evolution, if the number of particles in the *pbest* set
exceeds the specified capacity, the particle closest to the current particle is removed.
This strategy maintains the exploration of broader, potentially unknown areas in the

Fig. 8.6 Niche count

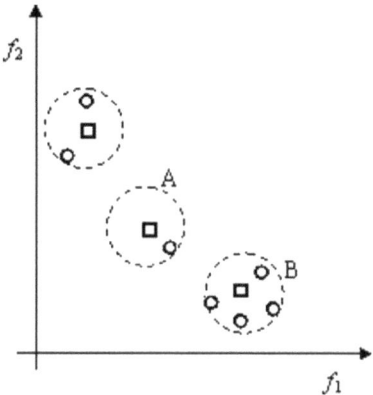

search space. In the later stages of evolution, however, the closest particle is retained to enhance the algorithm's convergence speed.

Of course, within "cluster-based methods", many algorithm improvements based on distance calculations have introduced distance-based concepts and techniques. Mostaghim [14] employed a σ-distance to partition the fitness space into multiple sub-spaces. Hsieh [28] implemented a fixed clustering radius for cluster centers. Particles at the center of this clustering radius are discarded.

(3) **Grid-based methods**

Coello [52, 53] introduced an innovative approach that goes beyond merely using external archives to store non-dominated solutions discovered during the search. This method employs a position-based strategy to partition the external archive into multiple hyperspaces (grids), with each particle's flight experience stored in a specific hyperspace. The adaptive grids, distributed uniformly along the *Pareto* front of solutions, play a crucial role in maintaining swarm diversity. All non-dominated solutions are stored within a hyperspace. The diversity index of hyperspace in the fitness space is equivalent to 1/10 of the contained number of particles. The grid size is finite, and the selection of *nbest* is based on particle density, with the selection probability proportional to the number of dominating solutions within that hyperspace. Therefore, particles in less dense grids have a higher preservation priority than those in denser grids. The position distribution-based selection method is divided into two steps. First, a roulette wheel method is used to select a hyperspace from the external archive grids, with the selection probability inversely proportional to the number of dominating particles in that space (Fig. 8.7). In Cagnina et al. [53] an elitist strategy is introduced for non-dominated solutions in the external archive, with this archive using a grid structure and uniform mutation probability. Sierra and Coello Coello [57] also applied the concept of grids in their design of external archive design.

As an improved archive structure, Beielstein [58] adopted an enhanced elitist mechanism and adaptive grid to partition the archive. It changes its own space size in the hyperspace at each iteration based on selection pressure and particle density. For particles in the archive, through the calculation of distance pairs in the fitness

Fig. 8.7 Coello grid

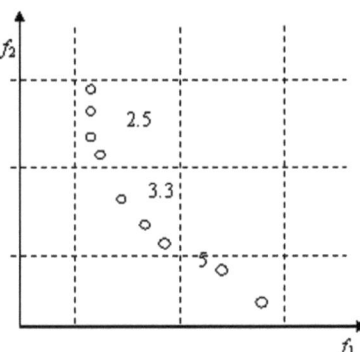

space, uniform selection probability and anti-clustering selection based on deletion probability are used to prevent excessive particle clustering. Only if the searched solution cannot be dominated by all solutions in the external archive, it preserved as a non-dominated solution. Based on fitness levels, new particles are allowed to replace non-dominated solutions in the archive, improving the particle distribution in the specified archive. Both *gBest* and *pBest* positions are selected from the external archive.

Leong [59] designed a crowdness indicator to monitor particle density in a grid. The crowdness indicator is defined as the number of particles in that grid, and it plays an important role in controlling the number of particles in the grid. Similar to the adaptive grid process, the adaptive local archive improves the distribution probability of each sub-swarm along the *Pareto* front. Sub-swarms are generated by a clustering algorithm based on the positions of optimal particles, while the selection of optimal particles adopts the roulette wheel method.

(4) **Cluster-based methods**

Mostaghim [32] employs a fixed-size storage space to preserve non-dominated solutions. The fitness space is partitioned into multiple subspaces using a distance-based method (Fig. 8.8). Each particle identifies the optimal guiding non-dominated solution within the minimum distance as its neighborhood best (*nbest*). A perturbation, based on random numbers relative to the current position, is incorporated into the position update formula. Clustering is utilized to retain superior elite particles, with inferior particles being removed when the storage space limit is reached. The σ-distance enhances particles' ability to converge directly towards the *Pareto* front. In subsequent research, Mostaghim [60] introduced a two-step optimization process, incorporating a multi-level partitioning method that iteratively divides the search space into subspaces. This approach imposes strict space limitations during the initial phase while allowing unrestricted space in later search stages.

Hsieh [28] also adopted clustering storage, in accordance with Ghosh et al. [32], to maintain diversity among currently identified non-dominated solution particles. Each particle is assigned a fixed clustering radius, with particles at the center of this radius being discarded (Fig. 8.9). *Hsieh* developed a solution exploration strategy (SES),

Fig. 8.8 σ-distance-based
partitioning of fitness space

Fig. 8.9 *Hsieh*'s diversity
preservation strategy for
clustering storage

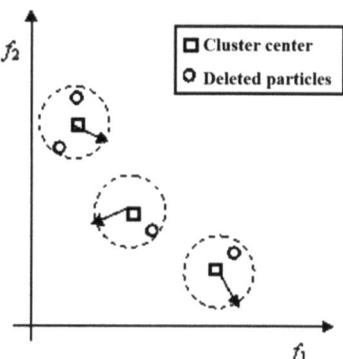

where particles are uniformly distributed around non-dominated solutions, and these
non-uniformly distributed solutions are used as references for local motion. Each non-
dominated solution in archive serves as a potential *gbest*, ensuring a wide distribution
of particles across the entire region. Perturbation operations are integrated into the
update equations to facilitate particles' escape from local minima and exploration of
unknown regions.

In contrast to using a single archive to store the selected *gbest* individuals, Abido
[51] employs two separate archives to keep track of *gbest* and *pbest*. Up to the current
iteration, all non-dominated solutions discovered by the particles are stored in a
global non-dominated set, while the non-dominated solutions obtained by individual
particles are kept in a local non-dominated set. A hierarchical clustering algorithm,
as utilized in SPEA, is applied to ensure that the number of individuals in each
set does not exceed its maximum capacity. If the number of individuals in either
non-dominated solution set exceeds its maximum capacity, the size of the local or
global non-dominated set is reduced based on average clustering. Each particle in
the swarm selects the nearest non-dominated solution from both the global and local
non-dominated sets according to the principle of proximity, thus designate them
as *gbest* and *pbest*, respectively. Leong [59] uses a clustering algorithm based on
optimal particle positions to partition sub-swarms.

Praveen [29] also employs a method with a fixed archive size but uses a classification approach based on ε-fuzzy dominance. At the current iteration and across all historical positions, the optimal N non-dominated solutions are selected using the fuzzy dominance and stored in an external archive. In each iteration, in addition to the update equations for positions and velocities, the K-means algorithm is used to divide the swarm into smaller sub-swarms, while the Nelder–Mead method (downhill simplex method) further refines local searches within each clustered sub-swarm. In Pulido and Coello Coello [61], *Johnson*'s clustering algorithm is employed to select optimal particles to guide the swarm during the search process.

(5) Hierarchical ranking

Various hierarchical and ranking methods are employed to classify or rank the non-dominated solutions in the *Pareto* set, followed by the selection of optimal individuals.

(1) *Pareto* ranking

Non-dominated particles are ranked based on *Pareto* optimality (*Pareto* ranking). During each iteration, particles are either added to or removed from the set of leaders (SOL) [12] based on comparisons of their non-dominance ranks. In unconstrained optimization, ranking is determined by comparing the average ranks of objective function values, while in constrained optimization, it is based on comparing the average ranks of the objective function relative to the constraints. Within the SOL, a strategy is employed to select optimal particles (*nbest*) using a roulette wheel method and a probability proportional to the number of particles within a unit radius and the crouwed radius. This approach increases the likelihood of selecting optimal individuals that are surrounded by fewer particles, thereby enhancing the swarm's ability to explore new areas while maintaining diversity. The combination of *Pareto* ranking and selection strategy allows the SOL structure to search along the *Pareto* front. For instance, Mandal and Mukhopadhyay [62] employed a MOPSO algorithm to identify non-redundant gene markers from microarray gene feature data for cancer characterization, incorporating non-dominated sorting as one criterion to determine whether updates.

(2) Dominated tree

To find a more effective method for selecting *gbest* compared to density-based selection methods, Fieldsend [63] employed a dominated tree in the external archive to choose an appropriate *Pareto* optimal solution as the *gbest*. The global non-dominated solutions in the external archive are sorted in a specific order and organized into a dominated tree structure. Additionally, *lbest* (*pbest*) particles are selected with equal probability from the local *Pareto*-optimal solution set, rather than all particles rely on a single unique optimal value. A perturbation is also introduced into the velocity update equation; the *gbest* for particles is calculated strictly within a certain area without any randomness, while the perturbation variable is a random number, similar to perturbations in evolutionary strategies.

(3) **Fitness sharing**

Salazar-Lechuga [64] considers both dominance and sharing of solutions within a specified archive capacity. Fitness sharing is used to evaluate the density of existing non-dominated solutions. A non-dominated solution that shares resources with other particles has its fitness value inversely related to the number and proximity of surrounding particles. A roulette wheel method is used to select particles (*Pareto* non-dominated solutions) from the archive based on fitness sharing. If the number of non-dominated solutions in the archive reaches its limit, particles with the worst fitness will be replaced by new ones.

(6) **Sub-swarm-based methods**

Reyes-Sierra [57] leverages the advantages of co-evolutionary methods by integrating diversity strategies into the basic PSO to tackle MOPs. Particles are divided into different sub-swarms, enabling collaboration and competition among them while adjusting the size of their sub-spaces based on the distribution along the current *Pareto* front. The study employs an adaptive grid to store non-dominated solutions, enhancing the uniform distribution of particles along the *Pareto* front. Additionally, clustering is utilized to analyze potential regions, and GA operations such as crossover and mutation are introduced, applying three types of mutation operations to the particles in the swarm (no mutation, uniform mutation, and non-uniform rate mutation).

Based on the research in Mostaghim and Teich [14, 60] also adopts a sub-swarm optimization approach. During the initial search phase, the swarm is divided into multiple sub-swarms based on σ-distance, with no restrictions on the size of the external archive for non-dominated solutions, allowing for the retention of as many non-dominated solutions discovered by the sub-swarms as possible.

In Pulido and Coello Coello [61], each sub-swarm employs its own PSO algorithm, with certain particles within each sub-swarm exchanging information. *Johnson*'s clustering algorithm is used to select optimal particles to guide the swarm performing search. Similar to the adaptive grid process, Leong and Yen [59] introduced an adaptive local archive mechanism into MOPSO and utilized clustering algorithms for classification to improve the distribution of *Pareto* front sub-sections relevant to each sub-swarm.

Sun et al. [65] propose a MOPSO algorithm where swarms have distinct roles. Some swarms dynamically search for improved *Pareto* fronts along the existing *Pareto* front, while others explore areas farther from it, thereby enhancing the algorithm's global search capability. Liang et al. [66] introduce a MOPSO algorithm that includes n swarms, where individuals are randomly assigned to n regions based on a randomly selected objective function. Individuals then dynamically select their historical *pbest* and *nbest* within their respective sub-regions.

8.2.2.3   Extended Methods

In practical applications, as the evaluation number of the fitness function increases, the associated computational cost becomes significantly higher. This is true not only for real-world problems but also the mathematical formulation of multi-objective problems becomes complex. Frequent evaluations of the fitness function can consume a substantial amount of time, leading to low efficiency in the optimization process. Therefore, to reduce the computational burden of fitness evaluations and enhance optimization performance, various extended improvement methods have been introduced in EA, such as methods for extending fitness evaluations. Fitness inheritance and fitness approximation are two typical extended methods. In recent years, these two methods have also been applied to MOPSOs.

(1)   **Fitness inheritance**

Fitness Inheritance refers to calculating the fitness values of particles in the next iteration based on the information from particles in the current swarm using specific approaches. Fitness inheritance can be divided into two categories [67]: algebraic mean or weighted mean of fitness values of two parent particles. Parent particles refer to current particles: *pbest*, *gbest*, or non-dominated solutions. Through various selection methods for parents and different weighting approaches, Reyes-Sierra [57, 67] proposed 15 methods for fitness inheritance. The *linear combination based on distance* (LCBD) method calculates new fitness values for particles through linear weighting based on their positions. It has three different forms depending on the selection method of parent particles. The *flight formula on objective space* (FFOS) method has a mathematical formulation that is formally similar to the update equations for particle velocities and positions, and also considers three different forms for evaluating the fitness values of different particles. The *combination using flight factors* method includes two types of implementation: linear weighting and nonlinear weighting, and there are six distinct equations for calculating fitness values. The expressions for linear and nonlinear weighting are similar, with the main difference being whether the fitness values of parent particles are combined linearly or nonlinearly. Research on the proportion of inheritance for fitness values indicates that [68], without compromising the quality of the obtained *Pareto* front, methods based on fitness inheritance can reduce the computational load by 32%. Only if more than 50% of evaluations of the fitness function are reduced, the optimization performance of the algorithm is significantly affected.

(2)   **Fitness approximation**

Fitness approximation primarily involves estimating a particle's current fitness based on the fitness values of neighboring particles from previous iterations. The simplest approach assigns the new particle the same fitness value as the nearest particle, which may appear similar to fitness inheritance. However, this method is actually derived from fitting models of the fitness landscape, such as polynomial fitting, neural networks, and interpolation regression.

Yapicioglu [69] applied the fitting of fitness values in his work. PSO utilizes an existing set of non-dominated solutions that have already been identified and then constructs a generalized regression neural network (GRNN) using these non-dominated solutions. This neural network can generate a considerable number of non-dominated solution sets.

8.3 Density Measurement and Diversity Maintenance

Unlike single-objective-based optimization problems that seek a unique optimal value, the primary task of MOPs is to obtain the *Pareto*-optimal solution set. Additionally, during the optimization process. All particle updates rely on the non-dominated solutions that have already been identified in the archive. Early studies on MOPSOs [25, 27] did not address the issue of maintaining diversity. However, the distribution of solutions in the fitness space is crucial; a denser and more uniform distribution of particles along the *Pareto* front is desirable. Therefore, maintaining swarm diversity is a key concern. One approach to achieving this is to control the search strategy for non-dominated solutions, preventing particles from clustering too closely around optimal particles. This method has been mentioned in earlier classifications of methods for MOPs. Another approach involves enhancing PSO's update equations and selection strategies by incorporating operations from EA, GA, or other optimization techniques.

Ho [70] introduced operations with random characteristics to improve the update equations while considering factors such as individual emotions, social experiences, velocity reversals, and directional mutations. He also introduced "craziness variables," which are perturbation operations with random magnitudes designed to maintain diversity among particles in the optimization algorithm. In addition to these four random parameters, an age variable for solutions stored in the archive is introduced to further enhance algorithm diversity. Solutions are selected based on their age variables and fitness weights using roulette wheel selection.

Although the mechanism of roulette wheel selection is designed to ensure that all *Pareto* individuals have the same selection probability, in practice, only a few densely populated regions of non-dominated solutions receive a higher selection probability. This bias is detrimental to maintaining swarm diversity. In order to address this issue, methods that combine roulette wheel selection with the distribution density of particles in the archive have been employed to improve this selection tendency.

The unidirectional information transfer in PSO and the broadcast-style information-sharing structure under the classical PSO social model represent unique forms of information interaction that differentiate PSO from GA. While this information transfer method allows PSO to converge rapidly, it also presents challenges for maintaining diversity. As a result, various social (information) topologies have been introduced in enhancements to PSO, along with operations commonly used in EA and GA, such as different forms of mutation and crossover operations, aiming at altering patterns of information transfer. Perturbation operations, similar to mutations in evolutionary strategies, involve randomly selecting position vectors (or velocity vectors) of particles and adding Gaussian white noise. This helps particles escape local minima and explore other unknown areas. Qu et al. [71] introduced crossover operations into MOPSO to enhance its search capabilities for multi-objective optimization. Zhang et al. [72] combined single-particle mutation from GA with PSO algorithms for co-evolution to solve MOPs.

Xu et al. [73] combined DE with PSO to tackle issues regarding environmental and economic adjustment. Hao and Qin [74] incorporated immune algorithms into PSO to address problems about robotic path planning.

8.4 Performance Metrics

Unlike the performance evaluation methods used for single-objective-based optimization problems, defining performance in MOPs is inherently complex. Typically, at least the following metrics should be considered [75]:

(1) The distance between the non-dominated solution set and the *Pareto* front;
(2) The distribution of the non-dominated solution set;
(3) The extent to which the non-dominated solution set covers the *Pareto* front.

In the study of MOPs using EA, a variety of evaluation methods have been developed [76–78]. The metrics employed in PSO primarily fall into the following categories:

(1) **Number of non-dominated solutions**. The performance of the algorithm is evaluated by comparing the number of non-dominated solutions it identifies. This is the simplest way to quantitatively assess the performance. However, relying solely on the count of non-dominated solutions to judge the method effectiveness for MOPs can be misleading.
(2) **Generational distance (*GD*)**. This metric calculates the distance between the non-dominated solutions found in the current iteration and the *Pareto* front. It indicates how far the discovered non-dominated vectors are from the *Pareto* optimal solution set [29]. The equation is as follows:

$$GD = \frac{(\sum_{i=1}^{n'} d_i^p)^{1/p}}{n'}, \; d_i = \min_{j=1}^{n} \sqrt{\sum_{k=1}^{M} (f_k^i - f_k^j)^2} \qquad (8.21)$$

where n represents the number of known non-dominated solutions on the *Pareto* front; n' represents the number of non-dominated solutions currently found by the algorithm; M represents the number of objectives; f_k^i represents the fitness value of the i-th individual for the k-th objective; and d_i represents the *Euclidean* distance.

(3) **Inverted generational distance (*IGD*)**. This metric is commonly used to measure the distance between the current generation's *Pareto* front obtained

by the algorithm and the actual *Pareto* front, reflecting both the distance and cover content of the *Pareto* front to the actual *Pareto* front.

$$IGD = \frac{\sqrt{\sum_{i=1}^{n} d_i^2}}{n} \tag{8.22}$$

(4) **Spacing (S)**. This metric describes the distribution of solution vectors. Additionally, by comparing it with the actual *Pareto* front, it provides a measure of how well the non-dominated solution vectors identified by the algorithm cover that front

$$S = \sqrt{\frac{1}{n-1} \sum_{i=1}^{n} (\bar{d} - d_i)^2}, \bar{d} = \frac{1}{n} \sum_{i=1}^{n} d_i \tag{8.23}$$

(5) **Error ratio (ER)**. Proposed by *Veldhuizen*, this metric indicates the probability that a non-dominated vector found by the algorithm is not an element of the actual *Pareto* front. According to its equation, if the i-th solution is an element of the actual *Pareto* front, then $e_i = 0$; otherwise, $e_i = 1$.

$$ER = \frac{\sum_{i=1}^{n'} e_i}{n'} \tag{8.24}$$

(6) **Success rate (SCC)**. This metric measures the number of elements found by the algorithm that belong to the *Pareto* front.

$$SCC = \sum_{i=1}^{n'} s_i \tag{8.25}$$

(7) **Set coverage (SC)**. Also known as the C-metric, SC(X', X'') maps the two sets X' and X'' to the range of $[0, 1]$, representing the ratio of solutions in X' that dominate solutions in X''. It is calculated as follows:

$$SC(X', X'') \triangleq \frac{\left| \{a'' \in X''; \exists a' \in X' : a' \preceq a''\} \right|}{|X''|} \tag{8.26}$$

If all solutions in X' can dominate all solutions in X'', then $SC = 1$; conversely, if no solutions in X' can dominate any solutions in X'', then $SC = 0$. Due to the presence of undetermined dominance relationships between the two sets, it is generally necessary to compute both SC(X', X'') and SC(X', X') to gain a clearer understanding of the dominance relationships between them.

(8) **Diversity** (Δ). This metric describes the distribution range of the obtained solutions and is defined as follows:

$$\Delta = \frac{d_f + d_i + \sum_{i=1}^{n'-1} \left| d_i - \overline{d} \right|}{d_f + d_i + (s-1)\overline{d}}, \overline{d} = \frac{\sum_{i=1}^{n'-1} d_i}{s-1} \qquad (8.27)$$

where d_f and d_i represent the *Euclidean* distances between extreme values and boundary solutions within the non-dominated solutions.

(9) **Convergence metric** (γ): Assume the *Pareto*-optimal set for MOPs be known. By uniformly selecting some points along the *Pareto* front, the minimum distance between the solutions obtained by the algorithm and these points is calculated. The average of all these minimum distances is defined as the convergence metric γ. As shown in Fig. 8.10, a smaller value of γ indicates that the algorithm is better at approximating the *Pareto*-optimal solution set. When all obtained solutions exactly coincide with the selected points, $\gamma = 0$.

(10) **Maximum coverage (MS)**. This metric represents the maximum range determined by two solutions that are farthest apart within the obtained non-dominated solution set.

$$MS = \sqrt{\frac{1}{m} \sum_{k=1}^{m} \left\{ \frac{\min(f_k^{\max}, F_k^{\max}) - \max(f_k^{\min}, F_k^{\min})}{F_k^{\max} - F_k^{\min}} \right\}} \qquad (8.28)$$

$f_k^{\max}$ and $f_k^{\min}$ represent the maximum and minimum values of non-dominated solutions for the k-th objective, respectively, while $F_k^{\max}$ and $F_k^{\min}$ represent the

Fig. 8.10 Convergence

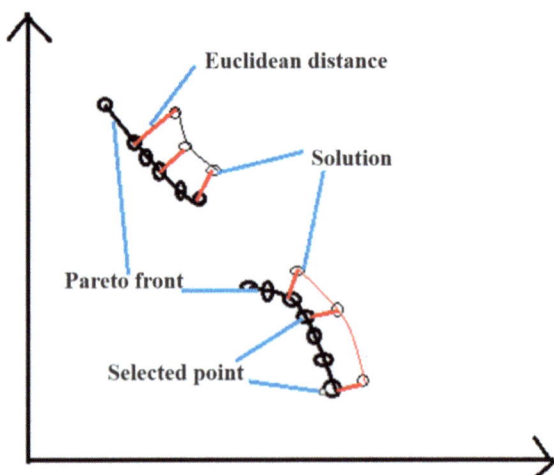

Table 8.3 Summary of metrics used in the referenced literature

Measurement method	References
Number of non-dominated solutions	[43, 27, 79, 25, 55]
Generation distance (GD)	[21, 24, 27, 28, 44, 47, 51, 53, 55, 80, 61]
Inverse generation distance (IGD)	[55, 67, 81]
Spacing (SP)	[24, 27, 44, 47, 53, 80, 61, 26, 49]
Error rate (ER)	[24, 27, 80, 61]
Success rate (SCC)	[55, 67, 81]
Set coverage (SC)	[24, 25, 44, 49, 81, 82]
Diversity (Δ)	[28, 55]
Maximum coverage (MS)	[28, 47, 53]

extreme values of the actual *Pareto* front. For bi-objective problems, this value indicates the *Euclidean* distance between two solutions. A larger matrix value signifies better optimization performance. Table. 8.3 summarizes the metrics used in the referenced literature.

8.5 Summary

This chapter begins by presenting the mathematical formulation of MOPs, along with benchmark functions, performance metrics, and classifications of these problems. This sets the stage for introducing MOPSO, which is categorized into different types based on its various characteristics.

The successful application of PSO in single-objective optimization has driven research into its use in multi-objective optimization, leading to the emergence of various MOPSO variants. Based on the different "decision-search" approaches to the problem, this article reviews and compares the improvement methods of MOPSO from both a priori and a posterior perspectives. Priori methods rely on prior knowledge about the objective functions, where the evaluation and selection order of these functions significantly influence the algorithm's outcomes. In contrast, posteriori methods primarily focus on three aspects: the non-dominated set, optimal particles, and diversity. Constructing the non-dominated set requires careful consideration of capacity control to reduce computational costs. This includes strategies for adding and removing non-dominated solutions as well as maintaining the non-dominated set, which is closely related to the algorithm's diversity strategies. Whether based on distance, grids, clustering, sub-swarms, or hierarchical ranking, these strategies serve as metrics and response strategies for swarm information. As a result, methods for measuring information and calculating spatial distances in information science have potential applications in this context. Additionally, various classification and sorting techniques from computational intelligence will provide powerful tools for constructing and maintaining non-dominated sets.

Subsequently, a summary of methods for fitting the fitness landscape is presented. In some respects, these methods are similar to progressive approaches. Neither of them simply execute optimization; instead, they incorporate the handling of the fitness landscape or the objectives to be optimized during optimization. To date, there is relatively little literature on this topic, especially concerning progressive MOPSO. However, this interaction will be a critical element in addressing practical multi-objective problems, making the solutions more rational. Additionally, this interactive intelligent search reflects *Kennedy*'s interpretation of PSO agents.

Any type of optimization problems, evaluating, judging, and measuring performance during the optimization process is essential from quantitative or qualitative perspectives. This paper summarizes nine performance metrics used in MOPSO, and each evaluating the problem from different perspectives. The discussed methods are derived from performance evaluation research in EA applied to multi-objective optimization. On the one hand, there are numerous performance metrics used in EA that are worth considering; on the other hand, algorithm evaluation must comprehensively balance computational cost and performance. Many optimization problems conduct confirmatory experiments under the assumption that the *Pareto* solution set is known. However, when the *Pareto* solution set is unknown, a comprehensive assessment of performance and computational cost becomes crucial.

References

1. Zou DL (2006) Key algorithms for QoS multicast routing research [D]. Dalian University of Technology
2. Schaffer JD (1985) Multiple objective optimization with vector evaluated genetic algorithms. In: Proceedings of the the 1st International conference on genetic algorithms, Lawrence Erlbsum
3. Fonseca CM, Fleming PJ (1993) Genetic algorithms for multi-objective optimization: formulation,discussion and generation. In: Proceedings of the the 5th International conference on genetic algorithms, San Mateo, California
4. Horn J, Nafpliotis N, Goldberg DE (1994) A niched Pareto genetic algorithms for multi-objective optimization. In: Proceedings of the the 1st IEEE congress on evolutionary computation, Piscataway
5. Srinivas N, Deb K (1994) Multi-objective optimization using nondominated sorting in genetic algorithms [J]. Evolut Comput 2(4):221–248
6. Zitzler E, Thiele L (2000) Multi-Objective evolutionary algorithms: a comparative case study and the strength Pareto approach [J]. IEEE Trans Evolut Comput 18(2):173–195
7. Deb K, Pratap A, Agarwal S et al (2002) A fast and elitist multiobjective genetic algorithm: NSGA-II [J]. IEEE Trans Evol Comput 6(2):182–197
8. Knowles JD, Corne DW (2000) Approximating the non—dominated front using the Pareto archived evolution strategy [J]. Evolut Comput 8(2):149–172(
9. Corne DW, Knowles JD, Oates MJ (2000) The Pareto-envelope based selection algorithm for multiobjective optimization[C]. In: Proceedings of the 6th International conference on parallel problem solving from nature, Berlin. pp 839–848

10. Zitzler E, Laumanns M, Thiele L (2001) Improving the strength Pareto evolutionary algorithm[C] [J]. In: SPEAZ: in evolutionary methods for design, optimization and control with applications to industrial problems, Athens, Greece. pp 95–100.
11. Parsopoulos KE, Vrahatis MN (2002) Particle swam optimization method in multiobjective problems[J]. In: Proceedings of the 2002 ACM symposium on applied computing. pp 603–607
12. Ray T, Liew KM (2002) A swarm metaphor for multiobjective design optimization [J]. Eng Optim 34:141–153
13. Lei DM, Wu ZM (2006) Pareto archive multi-objective particle swarm optimization [J]. Pattern Recogn Artif Intell 19(4):475–480
14. Mostaghim S, Teich J (2003) Strategies for finding good local guides in multi-objective particle swarm optimization (MOPSO). In: Proceedings of the 2003 IEEE Swarm intelligence symposium. SIS apos; 03
15. Hu X, Eberhart R (2002) Multiobjective optimization using dynamic neighborhood particle swam optimization. In: Proceedings of the 2002 congress on evolutionary computation. CEC'02 (Cat. No. 02TH8600). pp 1677–1681
16. Pulido GT, Coello CA (2004) Using clustering techniques to improve the performance of a multi-objective particle swarm optimizer[C]. In: Proceedings of the genetic and evolutionary computation conference GECCO. pp 225–237
17. Kejun W (2001) A new fuzzy genetic algorithm based on population diversity. In: Proceedings of 2001 International symposium on computational intelligence in robotics and automation. IEEE, Piscataway, NJ, USA
18. Coello CA, Lamont GB, Veldhuizen DA (2002) Evolutionary algorithms for solving multi-objective problems. Kluwer Academic Publishers, New York, USA
19. Engelbrecht AP (2009) Fundamentals of computational swarm intelligence. Tsinghua University Press, Beijing, pp 27–84
20. Zheng JH (2007) Multi-objective evolutionary algorithms and their applications[M] [J]. Science Press, Beijing
21. Lei DM, Yan XP (2009) Multi-objective intelligent optimization algorithms and their applications[M] [J]. Science Press, Beijing, p 38
22. Kirkpatrick S, Gelatt Jr CD, Vecchi MP (1983) Optimization by simulated annealin [J]. Science 220: 671–680
23. Coello Coello AC (2012) List of references on evolutionary multiobjective optimization[OL] [J]. https://delta.cs.cinvestav.mx/~ccoello/EMOO/emoopage.html
24. Li X (2004) Better spread and convergence: particle swarm multiobjective optimization using the maximum fitness function [J]. Lect Notes Comput Sci 3102:117–128
25. Moore J, Chapman R, Dozier G (2000) Multiobjective particle swarm optimization [J]. In: Proceedings of the 38th annual southeast regional conference, 2000, Clemson, South Carolina, USA. pp 56–57
26. Raquel CR, Naval PC (2005) An effective use of crowding distance in multiobjective particle swarm optimization [J]. In: Proceedings of the 2005 conference on genetic and evolutionary computation. ACM, Washington DC, USA, pp 257–264
27. Moore J, Chapman R (1999) Application of particle swarm to multiobjective optimization [J]. Department of Computer Science and Software Engineering, Auburn University
28. Hsieh S-T, Al E (2007) Cluster based solution exploration strategy for multiobjective particle swarm optimization [J]. In: Proceedings of the 25th conference on IASTED International Multi-Conference: artificial intelligence and applications 2007. ACTA Press, Innsbruck, Austria, pp 295–300
29. Koduru P, Das S, Welch SM (2007) Multi-objective hybrid PSO using μ-fuzzy dominance [J]. In: Proceedings of the 9th annual conference on genetic and evolutionary computation 2007. ACM, London, England, pp 853–860
30. Coello CA, Veldhuizen DAV, Lamont GB (2002) Evolutionary algorithms for solving multi-objective problems. Kluwer Academic Publishers
31. Hu X, Shi Y, Eberhart RC (2004) Recent advances in particle swarm. In: Proceedings of the Congress on evolutionary computation, CEC2004

32. Ghosh A, Chowdhury A, Sinha S, et al (2012) A genetic Lbest particle swarm optimizer with dynamically varying subswarm topology. In: Proceedings of the 2012 IEEE Congress on Evolutionary Computation (CEC). IEEE, Piscataway, NJ, USA

33. Koduru P, Das S, Welch SM (2007) Multi-objective hybrid PSO using N-fuzzy dominance. In: Proceedings of the 9th annual genetic and evolutionary computation conference, GECCO 2007, July 7, 2007–July 11, 2007. Association for Computing Machinery, London, United kingdom

34. Kuo RJ, Syu YJ, Chen Z-Y, et al (2012) Integration of particle swarm optimization and genetic algorithm for dynamic clustering [J]. Inform Sci 195:124–140

35. Jin X-L, Ma L-H, Wu T-J, et al (2007) Convergence analysis of the particle swarm optimization based on stochastic processes [J]. Acta Automat Sin 33(12):1263–1268

36. Liu DS, Al E (2006) On solving multiobjective bin packing problems using particle swarm optimization. In: IEEE congress on evolutionary computation

37. Coello Coello CA, Lechuga MS (2002) MOPSO: a proposal for multiple objective particle swarm optimization. In: Proceedings of the 2002 world congress on computational intelligence—WCCI'02. IEEE, Piscataway, NJ, USA

38. Fan HL, Zhong YC (2011) Two-subpopulation particle swarm optimization based on pheromone diffusion [J]. J Syst Simul 10:2125–2129

39. Xia WJ, Wu ZM (2005) Hybrid particle swarm optimization approach for multi-objective flexible job-shop scheduling problems [J]. Control Decis 20(2):137–141

40. Zhang XH, Zhou LX (2007) PID multi-object optimization design of PID controllers based on particle swarm algorithms [J]. J Appl Sci 25(4):392–396

41. Konstantinos EP, Michael NV (2002) Particle swarm optimization method in multiobjective problems. In: Proceedings of ACM symposium on applied computing

42. Liu W, Liang M (2006) A particle swarm optimization approach to a multi-objective reconfigurable machine tool design problem. In: IEEE congress on evolutionary computation

43. Marandi A, Al E (2006) Boolean particle swarm optimization and its application to the design of a dual-band dual-polarized planar antenna. In: IEEE congress on evolutionary computation. pp 3212–3218

44. Baumgartner U, Magele C, et al (2004) Pareto optimality and particle swarm optimization. IEEE Trans Magn 40(2):1172–1175

45. Zhang QF, Li H (2007) MOEA/D:A multiobjective evolutionary algorithm based on decomposition. IEEE Trans Evolut Comput 11(6):712–731

46. Li H, Zhang Q (2009) Multiobjective optimization problems with complicated pareto sets, MOEA/D and NSGA-II. IEEE Trans Evolut Comput 13(2):284–302

47. Peng W, Zhang QF (2008) A decomposition-based multi-objective particle swarm optimization algorithm for continuous optimization problems. In: 2008 IEEE International conference on granular computing. Piscataway, NJ, USA, pp 534–537

48. Zapotecas Martínez S, Coello Coello CA. A multi-objective particle swarm optimizer based on decomposition. In: Proceedings of the 13th annual conference on Genetic and evolutionary computation. Dublin, Ireland, pp 69–76

49. Carlo RR, Navaljr PC (2005) An effective use of crowding distance in multiobjective particle swarm optimization. In: Proceedings of the 2005 conference on genetic and evolutionary computation. ACM, Washington DC, USA, pp 257–264

50. Chavez HZ (2004) Artificial intelligence profits from biological lessons [J]. Distrib Syst IEEE 5(5):5

51. Abido MA (2007) Two-level of nondominated solutions approach to multiobjective particle swarm optimization. In: Proceedings of the 9th annual conference on Genetic and evolutionary computation 2007. ACM: London, England pp 726–733

52. Zhao B, Cao Y-J (2005) Multiple objective particle swarm optimization technique for economic load dispatch [J]. Zhejiang Univ SCI 6A(5):420–427

53. Cagnina L, ESQUIVEL S, Coello Coello AC (2005) A particle swarm optimizer for multi-objective optimization. J Comput Sci Technol 5(4):204–210

54. Cai ZQ, Huang H, Zheng ZH, et al (2009) Convergence improvement of particle swarm opti-mization based on the expanding attaining-state set [J]. J Huazhong Univ Sci Technol (Natural Science Edition) 6:44–47

55. Li X (2003) A non-dominated sorting particle swarm optimizer for multiobjective optimization [J]. In: Lecture notes in computer science. Springer-Verlag Heidelberg, Chicago, IL, USA

56. Xin B, Chen J, Zhang J, Fang H, Peng ZH (2012) Hybridizing differential evolution and particle swarm optimization to design powerful optimizers: a review and taxonomy. IEEE Trans Syst Man Cybernet Part C Appl Rev 42(5):744–767

57. Sierra MR, Coello Coello AC (2005) Improving PSO-based multi-objective optimization using crowding, mutation and e-dominance. Evolut Multi-Criter Optim 505–519

58. Bartz-Beielstein T, Limbourg P, Mehnen J, Schmitt K, Parsopoulos KE, Vrahatis MN (2003) Particle swarm optimizers for Pareto optimization with enhanced archiving techniques. In: The 2003 congress on evolutionary computation, apos;03, vol 3(8–12). pp 1780–1787

59. Leong W-F, YEN G G. Dynamic Population Size in PSO-based Multiobjective Optimization [J]. IEEE Congress on Evolutionary Computation, CEC 2006, 2006: p 1718–1725,

60. Mostaghim S, Teich J (2004) Covering Pareto-optimal fronts by subswarms in multi-objective particle swarm optimization. In: Proceedings of IEEE congress on evolutionary computation

61. Pulido GT, Coello Coello AC (2004) Using clustering techniques to improve the performance of a multi-objective particle swarm optimizer. In: Lecture notes in computer science. Seattle, WA, USA

62. Ting TO, Rao MVC, Loo CK (2006) A novel approach for unit commitment problem via an effective hybrid particle swarm optimization [J]. IEEE Trans Power Syst 21(1):411–418

63. Fieldsend J, Singh S (2002) A multi-objective algorithm based upon particle swarm optimi-sation, an efficient data structure and turbulence. In: The 00 UK workshop on computational intelligence. pp 34–44

64. Salazar-Lechuga M, Rowe JE (2005) Particle swarm optimization and fitness sharing to solve multi-objective optimization problems. In: The 2005 IEEE congress on evolutionary computation

65. Sun Y, Van Wyk BJ, Wang Z (2012) A new multi-swam multi-objective particle swam opti-mization based on Pareto front set. In: Huang DS, Gan Y, Gupta P, Gromiha M (eds.) Advanced intelligent computing theories and applications with aspects of artifical intelligence. Springer Berlin, Heidelberg, pp 203–210

66. Liang JJ, Qu BY, Suganthan PN, Niu B (2012) Dynamic multi-swarm particle swarm opti-mization for multi-objective optimization problems. In: Proceedings of the 2012 congress on evolutionary computation. pp 1–8

67. Reyes-Sierra M, Coello Coello CA (2005) A study of fitness inheritance and approximation techniques for multi-objective particle swarm optimization. In: The 2005 IEEE congress on evolutionary computation

68. Reyes-Sierra M, Coello Coello CA (2006) Dynamic fitness inheritance proportion for multi-objective particle swarm optimization. In: Proceedings of the 8th annual conference on Genetic and evolutionary computation 2006. ACM, Seattle, Washington, USA, pp 89–90

69. Yapicioglu H, Dozier G, et al (2006) Neural network enhancement of multiobjective evolutionary search. In: IEEE congress on evolutionary computation

70. Ho SL, Yang S, Ni G, Lo EW, Wong HC (2005) A particle swarm optimization-based method for multiobjective design optimizations. IEEE Trans Magn 41(5):1756–1759

71. Qu M, Gao YL, Jiang QY (2011) Multi-objective particle swarm optimization algorithm based on pareto neighborhood crossover operation. J Comput Appl 3(7):1789–1792

72. Zhang X, Dong H, Yang X, He J (2012) A mixed strategy multi-objective coevolutionary algorithm based on single-point mutation and particle swarm optimization. In: LI TR, Nguyen H, Wang GY, Grzymala-Busse J, Janicki R, Hassanien A, Yu H (eds) Rough sets and knowledge technology. Springer, Berlin, Heidelberg, pp 174–184

73. Xu LQ, Wu YL (2011) Multiobjective particle swarm optimization based on differential evolution for environmental/economic dispatch problem. J Xi'an Univ Technol 27(1):62–68

74. Hao W, Qin S (2011) Multi-objective path planning for space exploration robot based on chaos immune particle swarm optimization algorithm. In: Deng HP, Miao DQ, Lei JS, Wang F (eds.) Articial intelligence and computational intelligence[M]. Springer, Berlin, Heidelberg, pp 42–52

75. Zitzler E, Deb K, et al (2000) Comparison of multiobjective evolutionary algorithms: empirical results. Evolut Comput 8(2):173–195

76. Veldhuizen DAV, Lamont GB (1998) Multiobjective evolutionary algorithm research: a history and analysis [J]

77. Veldhuizen DAV, Lamont GB (1999) Multiobjective evolutionary algorithms: classifications, analyses, and new innovations

78. Hansen MP, Jaszkiewicz A (1998) Evaluating the quality of approximations to the non-dominated set [J]

79. Emma LB (2007) Optimising the flow of experiments to a robot scientist with multi-objective evolutionary algorithms [M]. In: Proceedings of the 2007 GECCO conference companion on Genetic and evolutionary computation 2007. ACM, London, United Kingdom, pp 2429–36

80. Coello CA, Pulido GT, Salazar-Lechuga M (2004) Handling multiobjectives with particle swarm optimization [J]. IEEE Trans Evol Comput 8(256–279)

81. Reyes-Sierra M, Coello C AC (2005) Fitness Inheritance in Multi-Objective Particle Swarm Optimization[C]. In: Proceedings of the IEEE swarm intelligence symposium. (SIS'05), Pasadena, California

82. Hu X, Eberhart RC, Shi Y (2003) Particle swarm with extended memory for multiobjective optimization[C]. In: Proceedings of the IEEE swarm intelligence symposium. (SIS 2003), Indianapolis, Indiana, USA

Chapter 9
Improvements to MOPSO

Similar to other EAs used for solving MOP, the design of applying PSO to MOP must consider several key aspects:

(1) To identify as many *Pareto*-optimal solutions as possible;
(2) To ensure that the identified *Pareto* front closely approximates the actual *Pareto* front;
(3) To achieve a well-distributed set of *Pareto*-optimal solutions, e.g., uniform distribution.
(4) To ensure that the algorithm converges quickly and can effectively address challenges in MOP, such as discontinuities in the *Pareto* front and *lbest*.

While GA shares information through chromosomes, the information-sharing mechanism in PSO is fundamentally different. In PSO, the entire swarm moves collectively toward optimal regions as a cohesive unit. Information is shared among individuals through *gbest*, *lbest*, or *nbest*, functioning like a black box where all individuals utilize the same model and search mechanism. This allows all particles to converge rapidly on optimal solutions, even within local neighborhoods [1]. Consequently, there are three additional considerations for designing PSO in MOP:

(1) How to select the *gbest* and *lbest* particles to ensure that the optimal particles must not only be *Pareto* optimal but also uniformly distributed along the *Pareto*-optimal boundary [2];
(2) How to maintain the non-dominated solutions that have already been identified in the swarm [3];
(3) Strategies for maintaining diversity to address diversity loss due to the pressure from information transfer [4].

The basic PSO algorithm cannot be directly applied to solve MOP. Therefore, we can either convert MOP into single-objective problems (as in traditional methods) or modify the basic PSO algorithm to enable it to tackle MOP challenges. When

© Beijing Institute of Technology Press 2025
F. Pan et al., *Particle Swarm Optimizer and Multi-Objective Optimization*,
https://doi.org/10.1007/978-981-95-3381-7_9

converting MOP into single-objective ones, the basic PSO algorithm can be effectively utilized. Here, we will introduce several common improved PSO algorithms for solving MOP, which are mainly based on dominance methods.

To test the following algorithms for MOPSO, we will select the unconstrained test functions ZDT1, ZDT2, ZDT3, and ZDT6, using GD and Δ as performance metrics.

9.1   CMOPSO

Coello and Lechuga proposed the MOPSO algorithm, which is based on an adaptive archive grid (CMOPSO), to address multi-objective problems. It is one of the earliest algorithms to widely utilize archiving methods [5, 6] and is also referred to as grid-based MOPSO. The core idea is to divide the objective space into several hypercubes and maintain the external archive by assessing the number of non-dominated solutions contained within each hypercube.

During each iteration, if the archive does not exceed a specified size, new non-dominated solutions will be added. If the archive is full, maintenance is performed based on the density of non-dominated solutions within the hypercubes; non-dominated solutions from high-density hypercubes are removed and replaced with those from low-density hypercubes, thereby ensuring swarm diversity.

In MOPSO, the selection of *gbest* and *pbest* solutions is also important. In CMOPSO, each particle's *gbest* is randomly selected from the external archive, with each particle having its own *gbest*. For the *pbest* of each particle, the algorithm employs the following strategy: a particle's optimal position will only change to a new solution if that solution dominates the original *pbest*.

The CMOPSO algorithm is described as follows:

Step 1: Create and initialize a swarm, setting the external archive *ex_archives* to an empty set;
Step 2: Evaluate all particles and add the resulting non-dominated solutions to the external archive;
Step 3: Maintain the external archive using the adaptive grid method;
Step 4: Select the *gbest* and *pbest* for each particle.
Step 5: Update the velocities and positions of the particles based on the swarm's velocity and position equations.
Step 6: Ensure that particles remain within the search space.
Step 7: Check if the termination conditions are met. If so, output the results and end the algorithm. If not, return to Step 2 and continue.

Regarding the selection of *pbest* in the adaptive grid-based MOPSO, only one selection strategy was mentioned earlier. Here, based on the characteristics of the algorithm, new strategies for modifying *pbest* and *gbest* are proposed:

(1) Selection of *pbest*: If a new solution dominates *pbest*, then *pbest* is updated to the new solution. If the new solution does not dominate *pbest*, then *pbest*

remains unchanged. If neither solution dominates the other, one is randomly selected from either the new solution or *pbest* to serve as the particle's *pbest* [7].

(2) Selection of *gbest*: Calculate the *Euclidean* distance between particle i in the swarm and the non-dominated solutions in the external archive. Select the m non-dominated solutions of the shortest distance, and choose the one with the minimum f_2 value as the *gbest* of particle i.

(3) In this framework, CMOPSO can be categorized into three types: CMOPSO_1 represents the original algorithm; CMOPSO_2 incorporates a modified *pbest* selection strategy; and CMOPSO_3 features a modified *gbest* selection strategy.

The three algorithms will be tested on ZDT1, ZDT2, ZDT3, and ZDT6, with parameters as outlined in Table 9.1.

MATLAB results are as follows:

① **Test results for ZDT1**

Figure 9.1 illustrates the curve of non-dominated solutions obtained by the algorithm compared with the *Pareto* front.

From Fig. 9.1, it is evident that all three algorithms effectively approximate the *Pareto* front of ZDT1, with CMOPSO-3 performing better than both CMOPSO-1 and CMOPSO-2.

Next, each algorithm runs 20 times to calculate *GD* and diversity metric $\triangle$, and the results are presented in Tables 9.2 and 9.3.

As shown in Table 9.2, *GD* for the CMOPSO-3 algorithm is smaller than that of CMOPSO-1 and CMOPSO-2 across minimum, maximum, variance, and average values. This indicates that the non-dominated solution set obtained by CMOPSO-3 is closer to the *Pareto* front and represents non-dominated solutions that are more accurate. The results for CMOPSO-2 are also excellent, while CMOPSO-1 performs relatively poorly.

Table 9.3 reveals that CMOPSO-2 has the best diversity metric, suggesting that the non-dominated solutions produced by this algorithm are more uniformly distributed along the *Pareto* front in objective space. Overall, CMOPSO-3 demonstrates superior

Table 9.1 Settings of CMOPSO parameters

Parameters	CMOPSO-1	CMOPSO-2	CMOPSO-3
ω	0.4	0.4	0.4
c_1, c_2	1.5	1.5	2
v_{max}	0.5	0.5	0.5
Popsize	50	50	50
Particle dimension	30	30	30
External archive size	100	100	100
Maximum number of iterations	3000	3000	3000

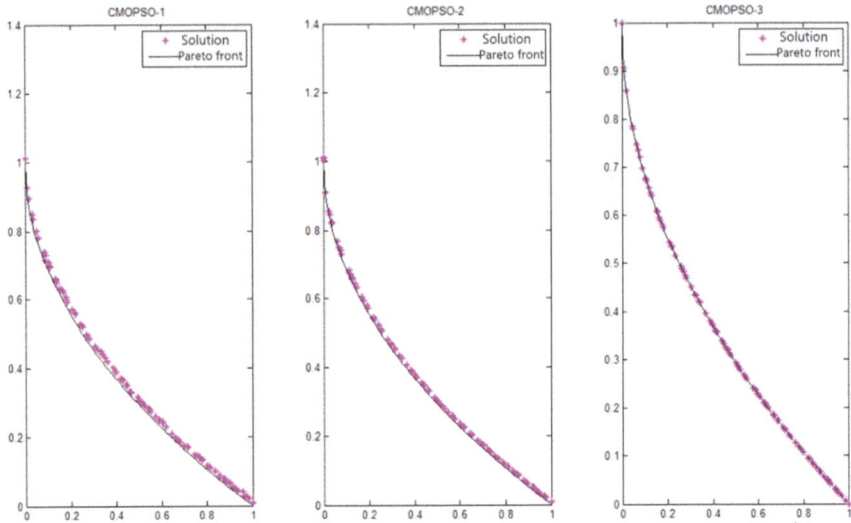

Fig. 9.1 Testing of the three algorithms on ZDT1

Table 9.2 *GD* for ZDT1 of CMOPSO1, CMOPSO2,CMOPSO3

Algorithm	Minimum value	Maximum value	Average value	Variance
CMOPSO-1	0.0019	0.0024	0.0023	1.4026×10^{-4}
CMOPSO-2	6.8796×10^{-4}	9.2746×10^{-4}	8.1281×10^{-4}	6.9352×10^{-5}
CMOPSO-3	9.1134×10^{-5}	2.0652×10^{-4}	1.3542×10^{-4}	3.6460×10^{-5}

Table 9.3 *Δ* for ZDT1 of CMOPSO1, CMOPSO2,CMOPSO3

Algorithm	Minimum value	Maximum value	Average value	Variance
CMOPSO-1	0.0178	0.0223	0.0214	0.0014
CMOPSO-2	0.0140	0.0181	0.0150	9.3540×10^{-4}
CMOPSO-3	0.0101	0.0191	0.0149	0.0021

performance; however, the diversity metric of CMOPSO-1 is inferior to those of the other two algorithms.

② **Test results for ZDT2**

Figure 9.2 shows the curve of non-dominated solutions obtained by the algorithm compared with the *Pareto* front, while Tables 9.4 and 9.5 shows *GD* and *Δ*, respectively.

The test results for the three algorithms on ZDT2 are fundamentally similar to those for ZDT1; however, regarding the diversity metric *Δ*, algorithm MOPSO-3 achieves the best results. In terms of both non-dominated solutions relative to the

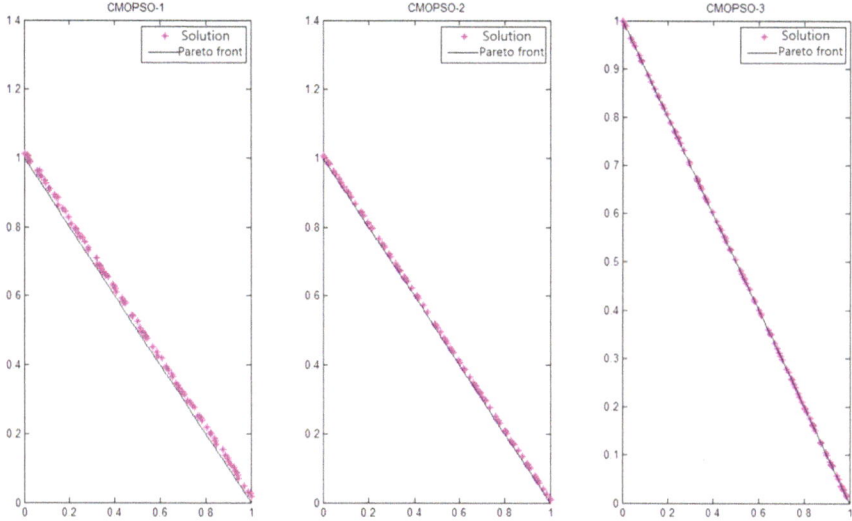

Fig. 9.2 Testing of the three algorithms on ZDT2

Table 9.4 *GD* for ZDT2 of CMOPSO1, CMOPSO2,CMOPSO3

Algorithm	Minimum value	Maximum value	Average value	Variance
CMOPSO-1	0.0021	0.0026	0.0023	1.3985×10^{-4}
CMOPSO-2	6.2653×10^{-4}	8.9633×10^{-4}	7.7189×10^{-4}	7.5307×10^{-5}
CMOPSO-3	7.6586×10^{-5}	9.3285×10^{-5}	8.2439×10^{-5}	4.3058×10^{-6}

Table 9.5 *Δ* for ZDT2 of CMOPSO1, CMOPSO2,CMOPSO3

Algorithm	Minimum value	Maximum value	Average value	Variance
CMOPSO-1	0.0201	0.0320	0.0261	0.0034
CMOPSO-2	0.0159	0.0288	0.0199	0.0033
CMOPSO-3	0.0101	0.0166	0.0109	0.0017

Pareto front curve and *GD*, as well as *Δ*, CMOPSO-3 outperforms the others, followed by MOPSO-2, while MOPSO-1 exhibits relatively poorer performance. This finding further supports previous analyses.

③ **Test results for ZDT3**

Figure 9.3 depicts the curve of non-dominated solutions obtained by the algorithm compared with the *Pareto* front. Table 9.6 shows the generational distance *GD*, while Table 9.7 illustrates the diversity metric *Δ*.

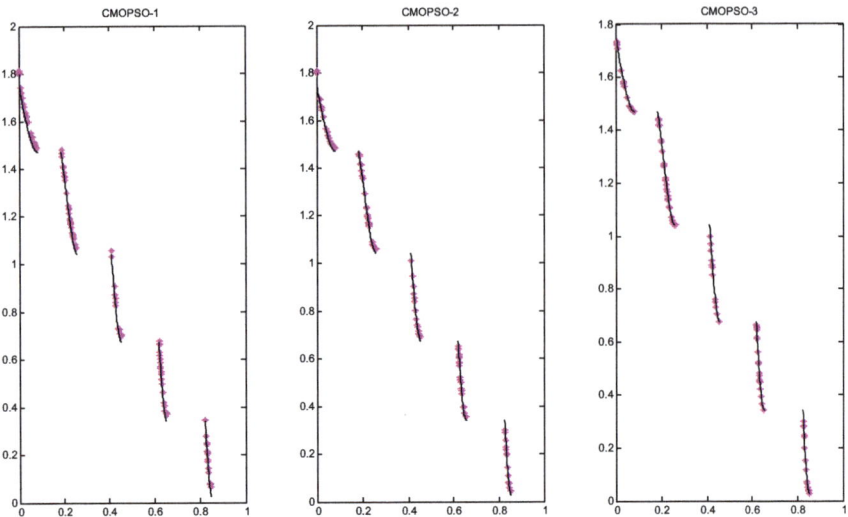

Fig. 9.3 Comparison of the three algorithms on ZDT3

Table 9.6 GD for ZDT3 of CMOPSO1, CMOPSO2,CMOPSO3

Algorithm	Minimum value	Maximum value	Average value	Variance
CMOPSO-1	0.0010	0.0015	0.0013	1.0751×10^{-4}
CMOPSO-2	5.6028×10^{-4}	8.2882×10^{-4}	6.9592×10^{-4}	5.5868×10^{-4}
CMOPSO-3	1.0805×10^{-4}	1.5222×10^{-4}	1.2883×10^{-4}	1.1496×10^{-5}

Table 9.7 Δ for ZDT3 of CMOPSO1, CMOPSO2,CMOPSO3

Algorithm	Minimum value	Maximum value	Average value	Variance
CMOPSO-1	0.0136	0.0165	0.0149	$7.7491*10^{-4}$
CMOPSO-2	0.0124	0.0136	0.0130	$2.7280*10^{-4}$
CMOPSO-3	0.0101	0.0113	0.0104	$3.4789*10^{-4}$

The graphical results and tabular data above lead to the same conclusions as before: the CMOPSO-3 algorithm exhibits superior performance, achieving the best generational distance *GD* and diversity metric *Δ* among the three algorithms.

④ **Test results for ZDT6**

When evaluating ZDT6, it's crucial to note that its objective space is not uniformly distributed. Along the global *Pareto* front, the *Pareto* solutions are non-uniformly distributed; solutions become sparser when closer to the *Pareto* front and denser when further away. As the three algorithms did not perform well on the 30-dimensional problem, a 10-dimensional problem was selected for testing. The results are as follows:

Figure 9.4 illustrates the curve of non-dominated solutions obtained by the algorithm compared with the *Pareto* front. Table 9.8 illustrates the generational distance *GD*, and Table 9.9 shows the diversity metric Δ.

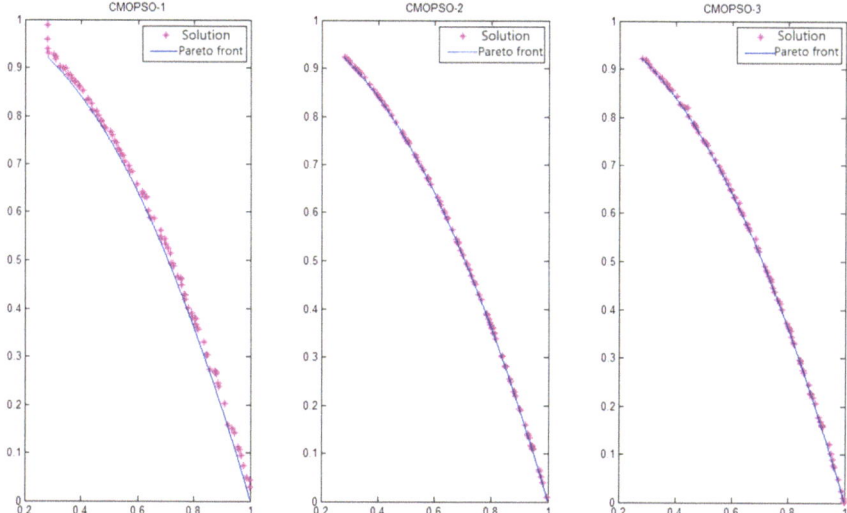

Fig. 9.4 Comparison of the three algorithms on ZDT6

Table 9.8 *GD* for ZDT6 of CMOPSO1, CMOPSO2,CMOPSO3

Algorithm	Minimum value	Maximum value	Average value	Variance
CMOPSO-1	7.1739×10^{-5}	0.0094	0.0025	0.0038
CMOPSO-2	1.2659×10^{-5}	7.3331×10^{-4}	1.8347×10^{-4}	1.9691×10^{-4}
CMOPSO-3	7.0085×10^{-5}	2.8303×10^{-4}	1.3060×10^{-4}	5.0730×10^{-5}

Table 9.9 Δ for ZDT6 of CMOPSO1, CMOPSO2,CMOPSO3

Algorithm	Minimum value	Maximum value	Average value	Variance
CMOPSO-1	0.0104	0.2998	0.0914	0.1270
CMOPSO-2	0.0109	0.0927	0.0524	0.0077
CMOPSO-3	0.0101	0.0135	0.0110	0.0012

Consistent with the results from the previous test, among the three algorithms tested on ZDT6, CMOPSO-3 exhibits superior performance, achieving the best generational distance *GD* and diversity metric Δ.

In summary, we can conclude that CMOPSO effectively approximates the *Pareto* front and produces a relatively uniform distribution of non-dominated solutions in the objective space. Its overall performance in solving MOP is sound. Furthermore, the improvements made to the selection strategies for *pbest* or *gbest* in the original CMOPSO algorithm have resulted in significant enhancements in its performance.

9.2 Multi-objective Comprehensive Learning Particle Swarm Optimization (MOCLPSO)

In single-objective particle swarm optimization algorithms, *Liang* et al. introduced an enhanced PSO algorithm known as the comprehensive learning particle swarm optimization (CLPSO) [7]. This algorithm implements a novel learning strategy that leverages the historical best positions of all particles to update particle velocities.

The fundamental concept of the CLPSO algorithm is as follows: particles in the swarm learn from three sources—the global best position (*gbest*), their own personal best position (*pbest*), and the historical best positions of other particles. For a particle with n dimensions, m dimensions learn from *gbest*, while the remaining $n-m$ dimensions randomly learn either from the particle's *pbest* or from the *pbest* of other particles.

The velocity learning can be described as follows:

For each particle i, generate a zero array a_i with length of n. Randomly generate a permutation r_c

of integers from 1 to n. Set $a(r_c(j))=1, j=1,2,...,m$; Randomly generate an array $b=rand(1,m)$;

$\quad$ *for* $i=1:N$ $\qquad$ (For each particle i in the swarm)

$\qquad$ *for* $j=1:n$ $\qquad$ (For each dimension of particle i)

$\qquad\qquad$ *if* $a(j)>P_e, v_{id}=\omega*v_{id}+c*rand()*(gbest_d-x_{id})$; $\qquad$ (learning from *gbest*)

$\qquad\qquad$ *elseif* $\quad b_i(j)>P_l, v_{id}=\omega*v_{id}+c*rand()*(pbest_{id}-x_{id})$; (learning from own *pbest*)

$\qquad\qquad$ *else* $\quad v_{id}=w*v_{id}+c*rand()*(gbest_d-x_{id})$; $\qquad$ (learning from other particles'

pbest)

$\qquad\qquad$ *end if*

$\qquad$ *end for*

In MOCLPSO, two key parameters require setting:

Learning probability P_l: This parameter determines whether a particle's dimension learns from its own *pbest* or from the *pbest* of other particles.

Elite probability P_e: In CLPSO, determining the number of dimensions that should be learned from *gbest* (i.e., the value of m) is difficult. To address this, an elite probability is introduced, which determines the size of m.

Huang et al. adapted CLPSO to tackle MOP, i.e., the multi-objective comprehensive learning particle swarm optimization (MOCLPSO). The specific process for selecting *pbest* is as follows: If *pbest* dominates the new solution *x*, then *pbest* remains unchanged; if *x* dominates *pbest*, then *pbest* is updated to *x*; if *pbest* and *x* are mutually non-dominating, one is randomly chosen from them as *pbest*. For selecting *gbest*, two approaches are employed: (1) Random selection; (2) Selecting the non-dominated solution closest to each particle as that particle's *gbest*. The strategy for velocity update was also refined: instead of learning from other particles' *pbest*, it now learns from other non-dominated solutions in the external archive.

MOCLPSO is categorized into two variants: MOCLPSO-1 and MOCLPSO-2. MOCLPSO-1 is the original algorithm where *gbest* is randomly selected, and the velocity learns from *gbest*, its own *pbest*, and other particles' *pbest*. MOCLPSO-2 employs an improved strategy for *gbest* selection, and its velocity learns from the particle's own *gbest*, its own *pbest*, and other non-dominated solutions in the external archive. These two algorithms will be evaluated on ZDT1, ZDT2, ZDT3, and ZDT6, with the results compared with those obtained from the CMOPSO-3 algorithm discussed in the previous section. The selected parameters are presented in Table 9.10.

MATLAB results are as follows:

① **Test results for ZDT1**

The figure below illustrates the test results of three algorithms for ZDT1. The curve represents the non-dominated solutions obtained by the algorithms compared to the *Pareto* front. The left graph shows the result of CMOPSO-3, the middle graph displays MOCLPSO-1, and the right graph presents MOCLPSO-2.

Figure 9.5 demonstrates that CMOPSO-3, MOCLPSO-1, and MOCLPSO-2 all effectively approximate the *Pareto* front. MOCLPSO-1 and MOCLPSO-2 exhibit a more uniform distribution along the *Pareto* front. Table 9.11 reveals that all three algorithms achieve very small *GD* for ZDT1. The diversity metric in Table 9.12 indicates

Table 9.10 Parameter settings of MOCLPSO

Parameters	CMOPSO-3	MOCLPSO-1	MOCLPSO-2
ω	0.4	0.4	0.4
c_1, c_2	2	1.5	2
v_{max}	0.5	0.5	0.5
popsize	50	50	50
Particle dimension	30	30	30
External archive size	100	100	100
Maximum number of iterations	3000	3000	3000
Number of grids	30	——	——
Learning probability P_l	——	0.1	0.1
Elite probability P_e	——	0.4	0.4

that MOCLPSO-1 and MOCLPSO-2 outperform CMOPSO-3, with MOCLPSO-2 yielding the best results.

② **Test results for ZDT2**

Figure 9.6 presents the test results of the three algorithms for ZDT2 (the curve of non-dominated solutions obtained by the algorithms compared to the *Pareto* front). Table 9.13 illustrates the generational distance *GD*, while Table 9.14 shows the diversity metric Δ.

The test results for ZDT2 closely resemble those for ZDT1. All three algorithms demonstrate a good approximation of the *Pareto* front for ZDT2. MOCLPSO-2 and

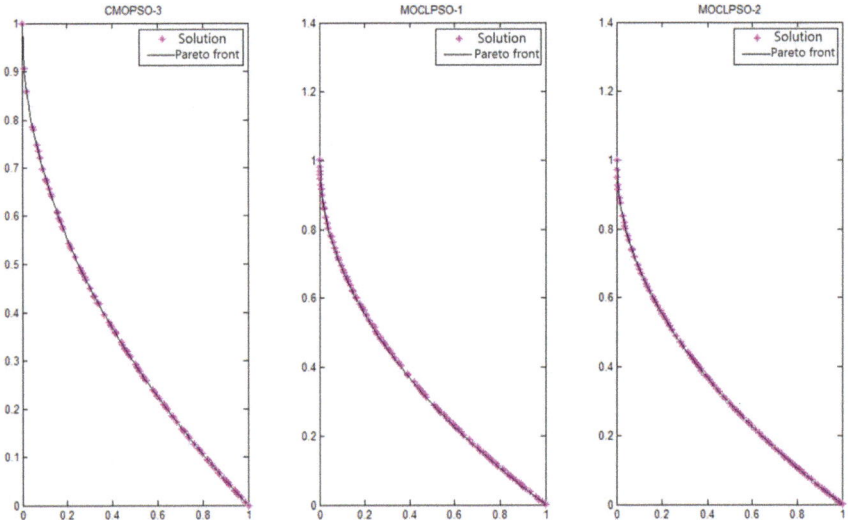

Fig. 9.5 Test results of three algorithms for ZDT1

Table 9.11 *GD* for ZDT1 of CMOPSO-3, MOCLPSO-1, MOCLPSO-2

Algorithm	Minimum value	Maximum value	Average value	Variance
CMOPSO-3	9.1134×10^{-5}	2.0652×10^{-4}	1.3542×10^{-4}	3.6460×10^{-5}
MOCLPSO-1	3.6931×10^{-4}	5.0349×10^{-4}	4.1437×10^{-4}	3.5559×10^{-5}
MOCLPSO-2	9.8422×10^{-5}	2.2901×10^{-4}	1.4000×10^{-4}	4.0578×10^{-5}

Table 9.12 Δ for ZDT1 of CMOPSO-3, MOCLPSO-1, MOCLPSO-2

Algorithm	Minimum value	Maximum value	Average value	Variance
CMOPSO-3	0.0101	0.0591	0.0149	0.0121
MOCLPSO-1	0.0103	0.0106	0.0104	6.8365×10^{-5}
MOCLPSO-2	0.0101	0.0101	0.0101	1.0136×10^{-5}

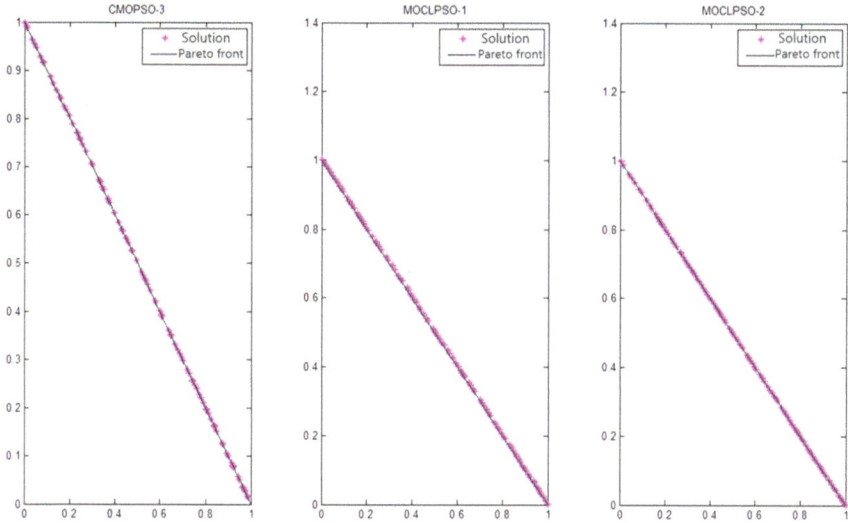

Fig. 9.6 Test results of three algorithms for ZDT2

Table 9.13 GD for ZDT2 of CMOPSO-3, MOCLPSO-1, MOCLPSO-2

Algorithm	Minimum value	Maximum value	Average value	Variance
CMOPSO-3	7.6586×10^{-5}	9.3285×10^{-5}	8.2439×10^{-5}	4.3058×10^{-6}
MOCLPSO-1	3.1662×10^{-4}	4.5784×10^{-4}	3.8587×10^{-4}	3.3127×10^{-5}
MOCLPSO-2	7.3256×10^{-5}	8.8383×10^{-5}	8.0944×10^{-5}	3.2715×10^{-6}

Table 9.14 Δ for ZDT2 of CMOPSO-3, MOCLPSO-1, MOCLPSO-2

Algorithm	Minimum value	Maximum value	Average value	Variance
CMOPSO-3	0.0101	0.0166	0.0109	0.0017
MOCLPSO-1	0.0111	0.0145	0.0121	8.3061×10^{-4}
MOCLPSO-2	0.0101	0.0101	0.0101	9.8202×10^{-7}

CMOPSO-3 achieve better *GD* than MOCLPSO-1, although all algorithms perform well overall. Regarding the diversity metric for ZDT2, MOCLPSO-2 exhibits the best performance, followed by MOCLPSO-1, while CMOPSO-3 shows relatively inferior results.

③ **Test results for ZDT3**

Figure 9.7 illustrates the test results of the three algorithms for ZDT3 (the curve of non-dominated solutions obtained by the algorithms compared to the *Pareto* front). Table 9.15 shows the generational distance *GD*, while Table 9.16 illustrates the diversity metric Δ.

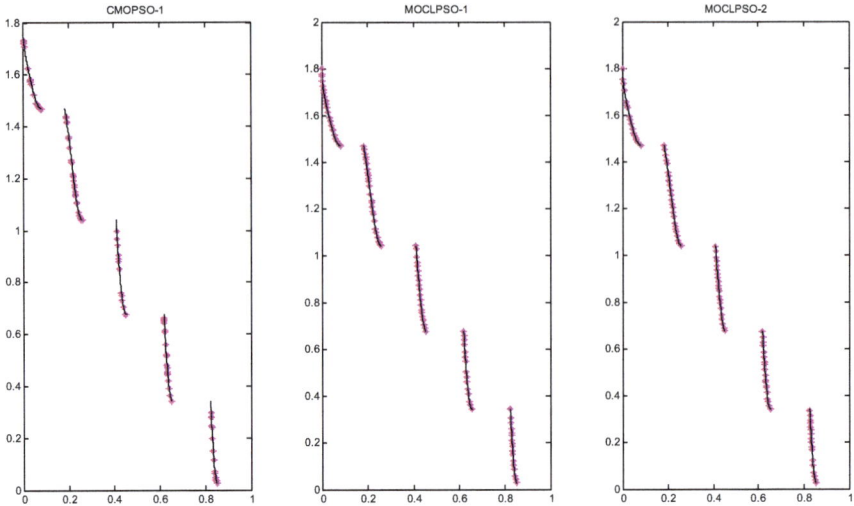

Fig. 9.7 Test results of three algorithms for ZDT3

Table 9.15 *GD* for ZDT3 of CMOPSO-3, MOCLPSO-1, MOCLPSO-2

Algorithm	Minimum value	Maximum value	Average value	Variance
CMOPSO-3	1.0805×10^{-4}	1.5222×10^{-4}	1.2883×10^{-4}	1.1496×10^{-5}
MOCLPSO-1	1.7511×10^{-4}	2.4711×10^{-4}	2.1328×10^{-4}	2.0871×10^{-5}
MOCLPSO-2	1.1378×10^{-4}	1.6798×10^{-4}	1.3729×10^{-4}	1.3213×10^{-4}

Table 9.16 *Δ* for ZDT3 of CMOPSO-3, MOCLPSO-1, MOCLPSO-2

Algorithm	Minimum value	Maximum value	Average value	Variance
CMOPSO-3	0.0101	0.0113	0.0104	3.4789×10^{-4}
MOCLPSO-1	0.0105	0.0108	0.0106	6.8678×10^{-5}
MOCLPSO-2	0.0101	0.0104	0.0102	7.3731×10^{-5}

Upon examination of the test results for ZDT3, we can draw conclusions consistent with the previous test results.

④ **Test results for ZDT6**

Figure 9.8 illustrates the test results of the three algorithms for ZDT6 (the curve of non-dominated solutions obtained by the algorithms compared to the *Pareto* front). Table 9.17 displays the generational distance *GD*, and Table 9.18 shows the diversity metric Δ.

Analysis of the test results for ZDT6 yields conclusions similar to those drawn from previous tests.

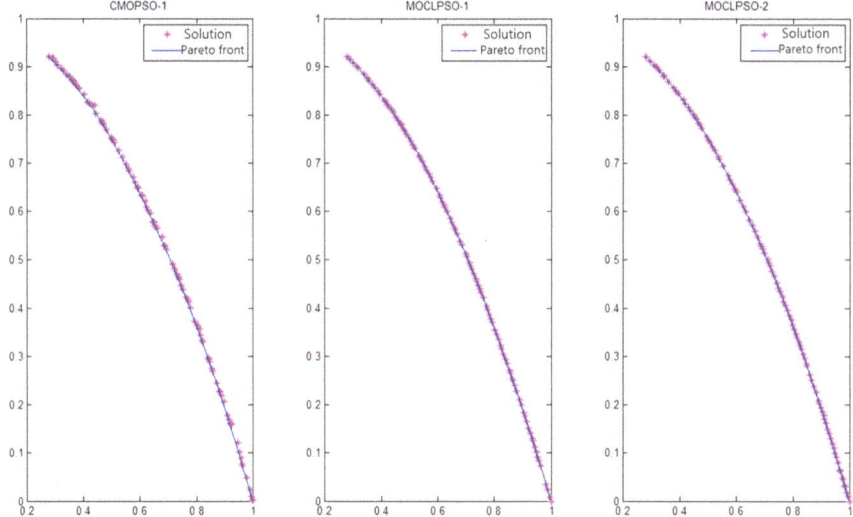

Fig. 9.8 Test results of three algorithms for ZDT6

Table 9.17 *GD* for ZDT6 of CMOPSO-3, MOCLPSO-1, MOCLPSO-2

Algorithm	Minimum value	Maximum value	Average value	Variance
CMOPSO-3	7.0085×10^{-5}	2.8303×10^{-4}	1.3060×10^{-4}	5.0730×10^{-5}
MOCLPSO-1	8.9577×10^{-5}	0.0014	2.7572×10^{-4}	3.9451×10^{-4}
MOCLPSO-2	0	1.6015×10^{-4}	7.6560×10^{-5}	2.7627×10^{-5}

Table 9.18 Δ for ZDT6 of CMOPSO-3, MOCLPSO-1, MOCLPSO-2

Algorithm	Minimum value	Maximum value	Average value	Variance
CMOPSO-3	0.0101	0.0135	0.0110	0.0012
MOCLPSO-1	0.0101	0.0127	0.0109	0.0010
MOCLPSO-2	0	0.0101	0.0097	0.0009

Based on the comprehensive analysis above, we can conclude that MOCLPSO demonstrates superior performance compared to CMOPSO. This superiority is particularly evident in terms of the diversity metric, where MOCLPSO significantly outperforms CMOPSO across all test functions.

9.3   Distance-Based Particle Swarm Optimization Algorithm (DISMOPSO)

In the previous discussion, we introduced two common algorithms for MOPSO (CMOPSO and MOCLPSO) and proposed improvements to them. While both algorithms aim to ensure diversity of solutions, they employ different selection strategies: CMOPSO uses a grid-based approach, while MOCLPSO utilizes a method based on crowding distance. Although these strategies can maintain solution diversity to some extent, they may also inadvertently limit the diversity of solutions obtained by the algorithms.

This section presents an improved algorithm derived from MOPSO, i.e., DISMOPSO, which employs a distance-based method for maintaining its external archive.

The selection strategy of DISMOPSO is as follows: in each iteration, the algorithm processes the non-dominated solutions in the external archive. For each non-dominated solution, it calculates the *Euclidean* distances to its neighboring non-dominated solutions (both left and right). If both distances are smaller than a predetermined value of *dis_1*, the solution is removed from the archive and the distance to the neighboring particles is recalculated. Figure 9.9 illustrates this process, where circular points represent non-dominated individuals and square points represent dominated individuals. If particle D is removed, the *Euclidean* distance between C and E (shown by the black line) must be recalculated. This selection strategy maintains the external archive by removing particles with poor diversity, even if the archive is not full. If particles with poor diversity (e.g., particle D) are not removed, the region containing particles C, D, E, and F (defined as region R) would be treated as a single area. Once during the random selection of *gbest*, particles in region R would have a higher probability of being chosen. This would lead to more non-dominated solutions being found in region R, thus resulting in decreased diversity of non-dominated solutions.

In this study, the critical distance *dis_1* is adaptively adjusted based on the non-dominated solutions using the following equation:

$$dis\_1 = \frac{\sum_{i=1}^{n-1} dis_i}{n+1} \tag{9.1}$$

where n represents the number of non-dominated solutions.

Fig. 9.9 New selection strategy

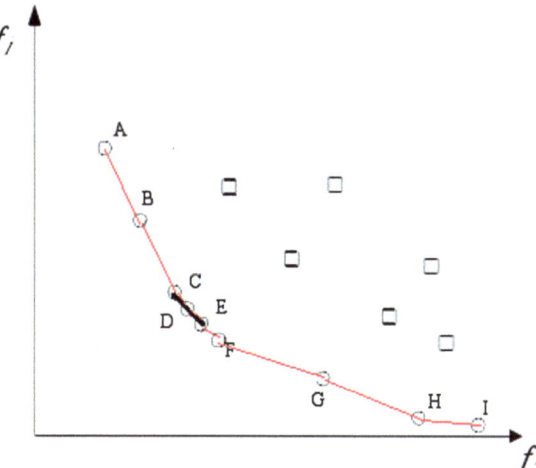

The process for maintaining the external archive is as follows:

(1) Sort all identified non-dominated solutions in ascending order based on the fitness value of a specific objective;
(2) Calculate the *Euclidean* distance between non-dominated solutions i and $i + 1$ in the objective space:

$$dis_i = \sqrt{\sum_{j=1}^{m}(f_i(j) - f_{i+1}(j))^2} \tag{9.2}$$

where m represents the number of objectives.

(3) For each calculated distance, check if it is less than the specified value of dis_1. If $dis_{i-1} < dis\_1$ and $dis_i < dis\_1$, remove non-dominated solution i, then recalculate the *Euclidean* distance between non-dominated solutions i-1 and $i + 1$ as the new dis_{i-1}.
(4) Determine if the number of non-dominated solutions exceeds the capacity of the external archive. If so, proceed to step 5; otherwise, skip to step 6;
(5) Let p be the number of non-dominated solutions exceeding the archive capacity, then:

---

for 1 to p

    [fr] = min(dis);

      if r == 1 then

          Remove the second non-dominated solution

          Calculate the distance between the first and third solutions as dis_1.

      elseif r is the last non-dominated solution, then

          Remove the second-to-last non-dominated solution

          Calculate the distance between the third-to-last and last solutions as the

          final distance value.

    else

        if $dis_{r-1} < dis_{r+1}$ then

           Remove the r-th non-dominated solution

           Calculate the distance between solutions r-1 and $r+1$ as dis_{r-1}.

        else

           Remove the $(r+1)$-th non-dominated solution

           Calculate the distance between solutions r and $r+2$ as dis_r.

        endif

      endif

endfor

---

(6) Return to the external archive.

The steps for DISMOPSO are as follows:

(1) Initialize a swarm and create an empty external archive (*ex_archives*);
(2) Evaluate all particles and add the resulting non-dominated solutions to the external archive;
(3) Maintain the external archive using the distance-based method described above;
(4) Select *gbest* and *pbest* for each particle;
(5) Update particle velocities and positions using the equations of standard PSO;
(6) Ensure all particles remain within the defined search space;
(7) Check termination conditions. If met, output results and terminate the algorithm. If not, return to step 2 and continue execution.

The DISMOPSO algorithm is tested on ZDT1, ZDT2, ZDT3, and ZDT6. The parameter settings are shown in Table 9.19. The results obtained are compared with those from the CMOPSO-3 and MOCLPSO-2.

Test Results for MATLAB:

For each algorithm, we conducted 20 runs with an external archive size of 20 and a swarm size of 100. Figures below illustrate the non-dominated solutions obtained by each algorithm compared to the *Pareto* front. In these figures, the left one represents CMOPSO-3 results, the middle one shows MOCLPSO-2 results, and the right one

Table 9.19 Parameter setting of DISMOPSO

Parameters	CMOPSO-3	MOCLPSO-2	DISMOPSO
ω	0.4	0.4	0.4
c_1, c_2	2	2	2
vmax	0.5	0.5	0.5
popsize	20	20	20
Particle dimension	30	30	30
External archive size	20	20	20
Maximum number of iterations	2000	2000	2000
Number of grids	30	——	——
Learning probability P_l	——	0.1	——
Elite probability P_e	——	0.4	——

displays the results of DISMOPSO. Figures 9.9, 9.10, 9.11, 9.12 and 9.13 represent the results of the three algorithms for the test functions.

Figures 9.10 and 9.11 clearly demonstrate the superiority of DISMOPSO. With an external archive size of 20, DISMOPSO not only approximates the *Pareto* front well but also achieves a highly uniform distribution along it. While solutions obtained via MOCLPSO approximate the *Pareto*-front well, they lack uniform distribution, indicating poor diversity. CMOPSO performs worst, with solutions reflecting only a portion of the *Pareto* front rather than covering it entirely. Figures 9.12 and 9.13 corroborate these findings.

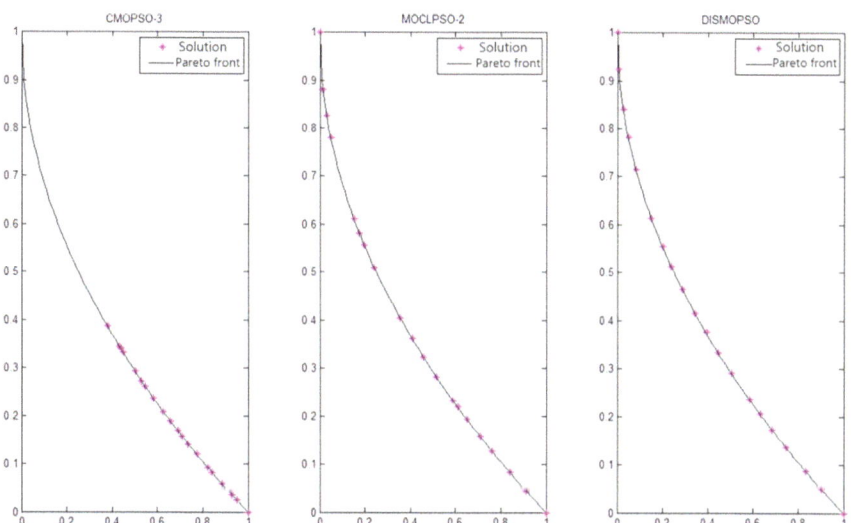

Fig. 9.10 Test for ZDT1 of CMOPSO-3, MOCLPSO-2, DISMOPSO

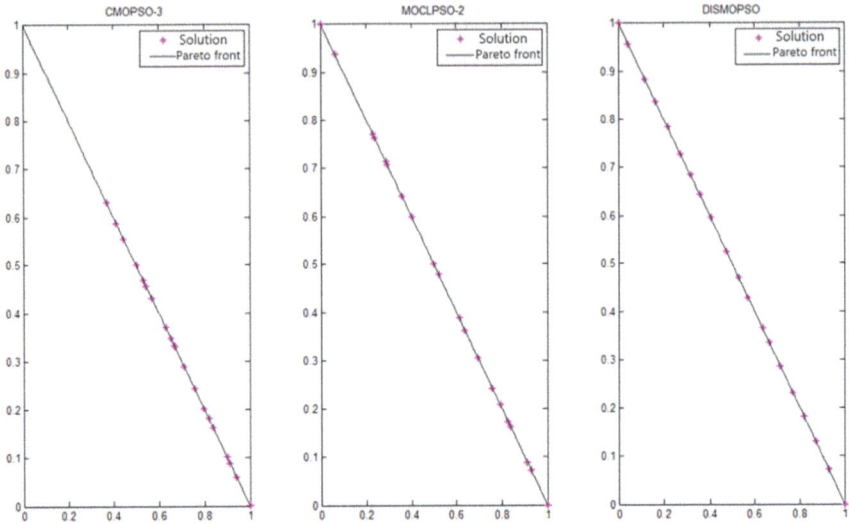

Fig. 9.11   Test for ZDT2 of CMOPSO-3, MOCLPSO-2, DISMOPSO

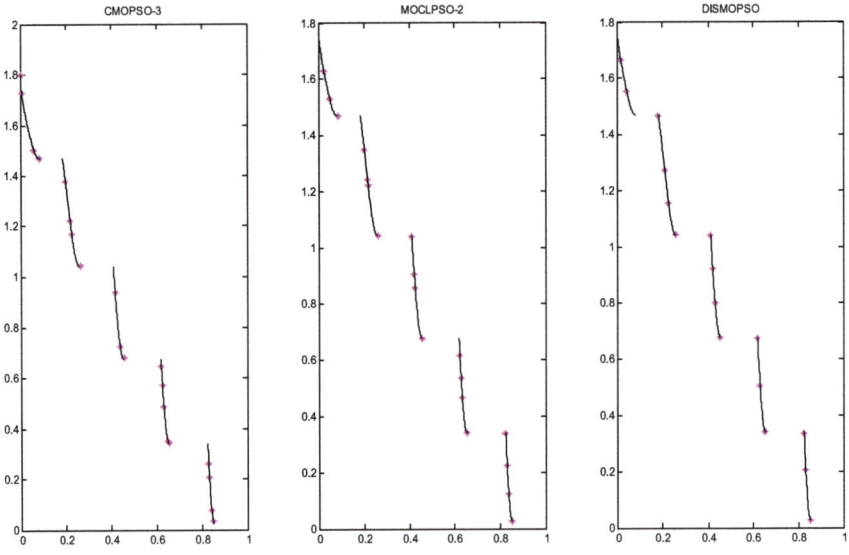

Fig. 9.12   Test for ZDT3 of CMOPSO-3, MOCLPSO-2, DISMOPSO

Table 9.20 shows that all three algorithms achieve good *GD*, typically in the order of 10^{-4} (with slightly worse results for ZDT6), indicating an effective approximation of the *Pareto* front. However, the results of CMOPSO, as seen in Figs. 9.10 and 9.11, reveal an incomplete approximation of the *Pareto* front, which is further confirmed by its Δ. Among the three algorithms, DISMOPSO achieves the best Δ, indicating the

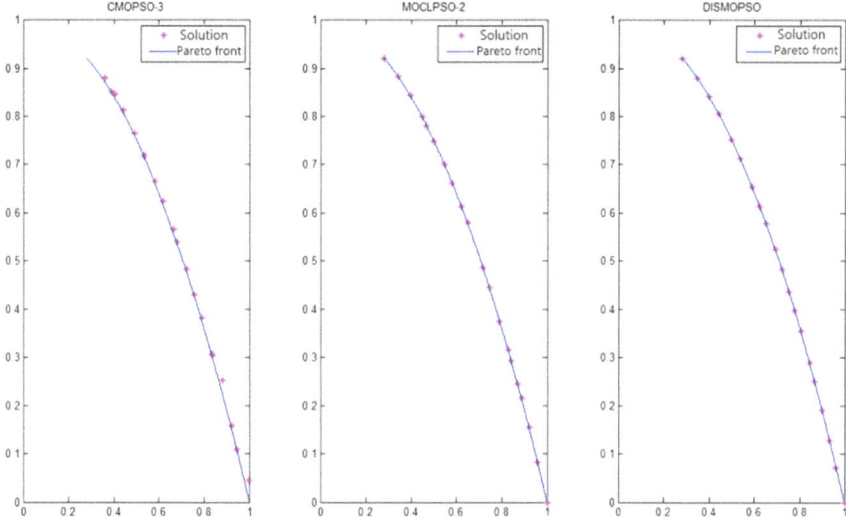

Fig. 9.13 Test for ZDT6 of CMOPSO-3, MOCLPSO-2, DISMOPSO

most uniform distribution of non-dominated solutions along the *Pareto* front. This results in a more comprehensive representation of the *Pareto* front, demonstrating the high quality of solutions obtained by DISMOPSO and validating its effectiveness.

The improved algorithm proposed in this section enhances the diversity of results compared to other MOPSOs, which facilitates the discovery of superior non-dominated solutions. However, there remains a gap in solution precision (as measured by *GD*) compared to CMOPSO and MOCLPSO. Addressing this precision issue is an area for future improvement.

9.4 Cooperative Multi-objective Particle Swarm Optimization (CMPSO)

Most MOPSO algorithms treat optimization problems as a whole, but the optimization objectives are often conflicting, making it difficult to assign fitness to individual particles. Zhan et al. [8] proposed a multi-population multi-objective (MPMO) strategy and applied it to MOPSO, resulting in Cooperative Multi-objective particle swarm optimization (CMPSO).

The basic idea of CMPSO is to divide the swarm into m sub-swarms based on the dimensionality m of the optimization objectives. Each sub-swarm selects its best particle (*gbest*) according to its assigned optimization objective. Individual particles learn from their personal best position (*pbest*), the sub-swarm's best individual (*gbest*), and particles from the external archive A.

For the i-th particle in the m-th sub-swarm, its velocity update equation is:

Table 9.20 Evaluation metrics for different algorithms (DISMOPSO, CMOPSO-1, MOCLPSO-2)

Algorithm		DISMOPSO		CMOPSO-1		MOCLPSO-2	
		GD	$\triangle$	GD	$\triangle$	GD	$\triangle$
ZDT1	Min	1.89×10^{-4}	0.119	1.36×10^{-4}	0.625	1.24×10^{-4}	0.244
	Max	9.57×10^{-4}	0.217	4.83×10^{-4}	0.753	6.77×10^{-4}	0.458
	Mean	4.00×10^{-4}	0.167	2.28×10^{-4}	0.677	2.63×10^{-4}	0.325
	Std	2.08×10^{-4}	0.026	9.57×10^{-5}	0.038	1.43×10^{-4}	0.053
ZDT2	Min	1.44×10^{-4}	0.123	1.74×10^{-4}	0.505	1.27×10^{-4}	0.222
	Max	7.68×10^{-4}	0.188	2.20×10^{-4}	0.650	2.22×10^{-4}	0.544
	Mean	3.40×10^{-4}	0.157	1.92×10^{-4}	0.578	1.74×10^{-4}	0.337
	Std	1.89×10^{-4}	0.022	1.26×10^{-5}	0.039	2.35×10^{-4}	0.077
ZDT3	Min	2.34×10^{-4}	0.123	2.16×10^{-4}	0.382	$2.21 \times 10{-4}$	0.175
	Max	3.26×10^{-4}	0.244	5.80×10^{-4}	0.699	5.85×10^{-4}	0.436
	Mean	2.76×10^{-4}	0.164	3.44×10^{-4}	0.536	3.09×10^{-4}	0.292
	Std	2.91×10^{-4}	0.033	1.26×10^{-4}	0.090	9.32×10^{-5}	0.062
ZDT6	Min	0.0001	0.101	0.0002	0.254	0.0002	0.163
	Max	0.0083	0.434	0.0297	0.595	0.0154	0.563
	Mean	0.0017	0.295	0.0040	0.402	0.0024	0.318
	Std	0.0059	0.089	0.0074	0.087	0.0044	0.101

$$v_{id}^m = \omega v_{id}^m + c_1 \cdot \text{rand}() \cdot (pbest_{id}^m - x_{id}^m) + c_2 \cdot \text{rand}() \cdot (gbest_{id}^m - x_{id}^m)$$
$$+c_3 \cdot \text{rand}() \cdot (A_{id}^m - x_{id}^m) \tag{9.3}$$

In CMPSO, the update rule for $pbest^m$ is as follows: if the m-th dimensional objective of x_i^m is smaller than that of $pBest_i^m$, then x_i^m replaces $pBest_i^m$. The update rule for $gbest^m$ is as follows: if the m-th dimensional objective of $pBest_i^m$ is smaller than that of $gbest^m$, then $pBest_i^m$ replaces $gbest^m$. A_i^m is randomly selected from the external archive.

The size of the external archive is set to a fixed value of NA. After each iteration, the $pbest$ of all particles from all sub-swarms are placed into the external archive. Dominance relationships are determined, and non-dominated solutions are retained. If the number of remaining solutions exceeds NA, then NA solutions are preserved based on density.

Now, CMPSO will be used to test ZDT1, ZDT2, ZDT3, and ZDT6 functions, and the results will be compared with those obtained from CMOPSO-3, MOCLPSO-2, and DISMOPSO. The parameters selected are shown in Table 9.21.

The test results are as follows:

The external archive size was set to 20, and the swarm size to 100. For CMPSO, two sub-swarms were established, each with a size of 50. Each algorithm was run 20 times. In the followings figure showing the non-dominated solutions obtained by the

Table 9.21 Parameter settings of CMPSO

Parameter	CMOPSO-3	MOCLPSO	DISMOPSO	CMPSO
ω	0.4	0.4	0.4	0.4
c_1	2	2	2	4.0/3
c_2	2	2	2	4.0/3
c_3	—	—	—	4.0/3
popsize	20	20	20	20
Particle dimension	30	30	30	30
Maximum number of iterations	2000	2000	2000	2000
Number of grids	30	—	—	—
Learning probability P_l	—	0.1	—	—
Elite probability P_e	—	0.4	—	—

algorithms compared to the curves of Pareto front, the first three figures represent the results of CMOPSO-3, MOCLPSO-2, and DISMOPSO respectively, while the fourth figure shows the results of CMPSO.

Figures 9.14, 9.15, 9.16 and 9.17 show the test results of the four algorithms for the test functions ZDT1, ZDT2, ZDT3, and ZDT6.

The evaluation metrics for the test results of each algorithm are summarized in Table 9.22.

Figures 9.14, 9.15, 9.16 and 9.17 clearly demonstrate that CMPSO achieves the closest approximation to the Pareto front, with solutions generally distributed across the entire Pareto optimal front. Other improved MOPSO algorithms show slight deviations from the Pareto optimal front in their non-dominated solutions. This indicates

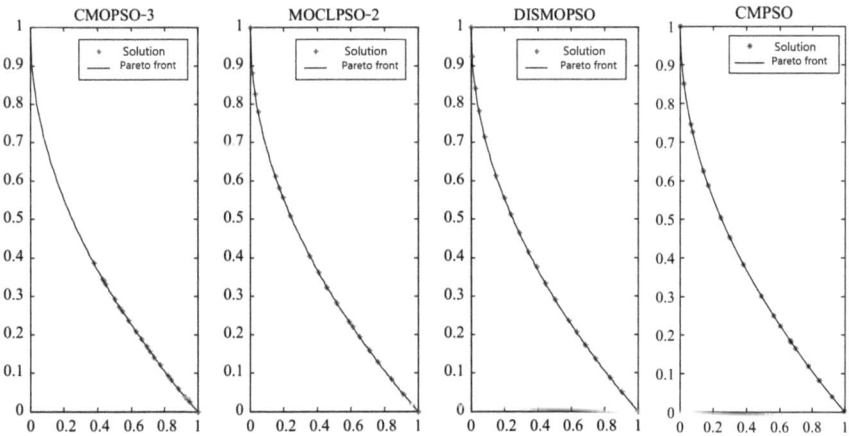

Fig. 9.14 Test for ZDT1 of CMOPSO-3, MOCLPSO-2, DISMOPSO, CMPSO

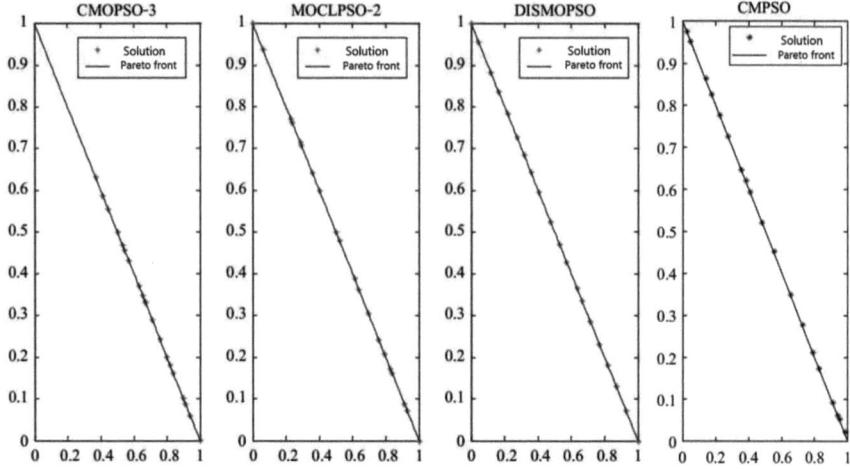

Fig. 9.15 Test for ZDT2 of CMOPSO-3, MOCLPSO-2, DISMOPSO, CMPSO

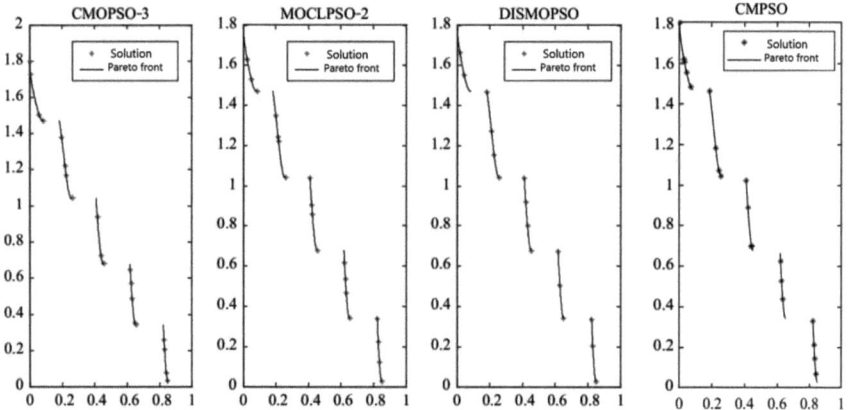

Fig. 9.16 Test for ZDT3 of CMOPSO-3, MOCLPSO-2, DISMOPSO, CMPSO

that the multi-swarm cooperative strategy enhances MOPSO's optimization capability. However, CMPSO's approach of particles learning from their sub-swarm's *best* may lead to an undesirable clustering of particles towards edge regions, potentially causing an imbalanced distribution of non-dominated solutions.

The data in Table 9.22 reveals that all four algorithms effectively approximate the Pareto front. CMPSO achieves the smallest value of *GD*, indicating its non-dominated solutions are closest to the *Pareto* front. This demonstrates CMPSO's superior performance in complex optimization problems. However, CMPSO's diversity metric (Δ) lags significantly behind DISMOPSO. This suggests that while CMPSO's solutions are closer to the *Pareto* front, their distribution lacks uniformity, potentially limiting

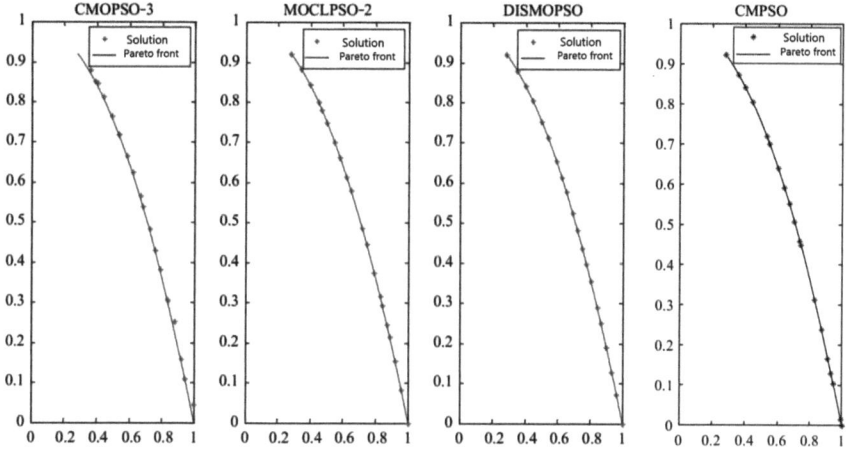

Fig. 9.17 Test for ZDT6 of CMOPSO-3, MOCLPSO-2, DISMOPSO, CMPSO

the comprehensive representation of the *Pareto* front. There is room for improvement in CMPSO's solution distribution.

However, the test functions chosen in this study are all bi-objective, and further research and analysis are needed for multi-objective functions (with more than two objectives). Therefore, applying the algorithm to multi-objective problems is part of future work. Moreover, multi-objective optimization problems often come with constraints, so handling these constraints in multi-objective optimization problems is also part of future work. Overall, the improved algorithm proposed in this paper performs better than CMOPSO and MOCLPSO algorithms to some extent. Future work will focus on more refined research of the algorithm to enable it to better solve multi-objective optimization problems.

9.5 Dynamic Heterogeneous Multi-swarm PSO (DHMSPSO)

PSO and most variants are primarily oriented toward solving static dimensional optimization problems. However, many real-world engineering problems exhibit dynamic dimensions. To address problems with dynamic dimensions, Chen [9] proposed the Dynamic Heterogeneous Multi-Swarm PSO (DHMSPSO) algorithm and demonstrated its application by optimizing the frequency assignment problem as an example.

The algorithm designs a dynamic variable structure, where the variable encoding consists of two parts: a static encoding component and a dynamic encoding component.

Table 9.22 Evaluation metrics for different algorithms (CMPSO, DISMOPSO, CMOPSO-1, MOCLPSO-2)

Algorithm		CMPSO		DISMOPSO		CMOPSO-1		MOCLPSO-2	
		GD	Δ	GD	Δ	GD	Δ	GD	Δ
ZDT1	Min	1.32×10^{-4}	0.371	1.89×10^{-4}	0.119	1.36×10^{-4}	0.625	1.24×10^{-4}	0.244
	Max	2.19×10^{-4}	0.514	9.57×10^{-4}	0.217	4.83×10^{-4}	0.753	6.77×10^{-4}	0.458
	Mean	1.71×10^{-4}	0.448	4.00×10^{-4}	0.167	2.28×10^{-4}	0.677	2.63×10^{-4}	0.325
	Std	1.22×10^{-4}	0.029	2.08×10^{-4}	0.026	9.57×10^{-5}	0.038	1.43×10^{-4}	0.053
ZDT2	Min	1.31×10^{-4}	0.378	1.44×10^{-4}	0.123	1.74×10^{-4}	0.505	1.27×10^{-4}	0.222
	Max	2.07×10^{-4}	0.592	7.68×10^{-4}	0.188	2.20×10^{-4}	0.650	2.22×10^{-4}	0.544
	Mean	1.73×10^{-4}	0.401	3.40×10^{-4}	0.157	1.92×10^{-4}	0.578	1.74×10^{-4}	0.337
	Std	1.44×10^{-4}	0.032	1.89×10^{-4}	0.022	1.26×10^{-5}	0.039	2.35×10^{-4}	0.077
ZDT3	Min	1.36×10^{-4}	0.344	2.34×10^{-4}	0.123	2.16×10^{-4}	0.382	2.21×10^{-4}	0.175
	Max	2.39×10^{-4}	0.506	3.26×10^{-4}	0.244	5.80×10^{-4}	0.699	5.85×10^{-4}	0.436
	Mean	1.79×10^{-4}	0.427	2.76×10^{-4}	0.164	3.44×10^{-4}	0.536	3.09×10^{-4}	0.292
	Std	1.83×10^{-4}	0.071	2.91×10^{-4}	0.033	1.26×10^{-4}	0.090	9.32×10^{-5}	0.062
ZDT6	Min	0.0001	0.231	0.0001	0.101	0.0002	0.254	0.0002	0.163
	Max	0.0079	0.526	0.0083	0.434	0.0297	0.595	0.0154	0.563
	Mean	0.0013	0.328	0.0017	0.295	0.0040	0.402	0.0024	0.318
	Std	0.0054	0.098	0.0059	0.089	0.0074	0.087	0.0044	0.101

$$x = (x_1, x_2, ..., x_s; x_{s+1}, x_{s+2}, ..., x_{s+d}) \tag{9.4}$$

During the iterative process, the algorithm selects particles as candidates for dimension mutation and appends a variable to the end of their encoding. If the fitness of the mutated individual exceeds that of the original individual, the mutation is considered successful. A mutated swarm is then derived from this new individual. Consequently, heterogeneous multi-swarms possessing different variable dimensions co-evolve to search for the optimal solution.

The algorithm flow of DHMSPSO is shown in Fig. 9.18.

In the frequency assignment problem (FAP), the entire operational process is divided into N_T consecutive operational phases. The frequency usage requirements

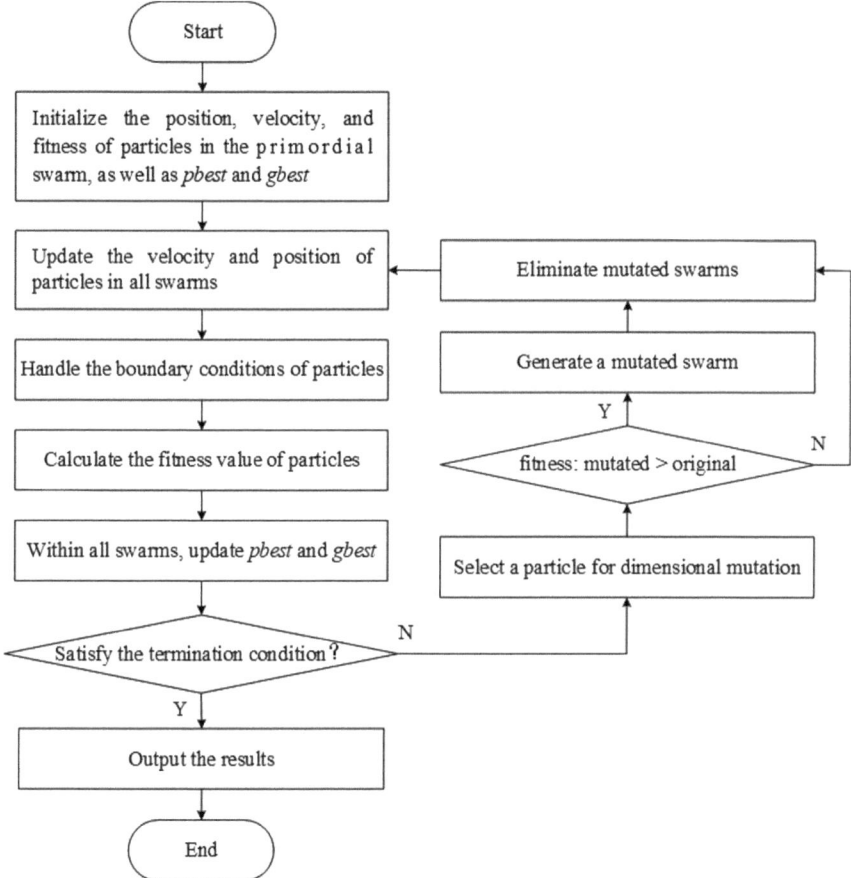

Fig. 9.18 Algorithm flow of DHMSPSO

of the frequency-dependent equipment in each operational phase are represented by "1" and "0".

The optimization objectives of the FAP are minimizing electromagnetic interference (EMI) and minimizing the number of frequency handoff operations, denoted as f_1 and f_2. Crucially, the second optimization objective is related to dynamic dimension, in this case DHMSPSO is applicable to optimizing the multi-objective FAP problem.

$$x_{i2} = x_{i1} + \beta \cdot r_g \tag{9.5}$$

For FAP, the dimension variation of particles is designed as follows. While retaining the optimization variables of the mutated particle, a frequency-dependent equipment is randomly selected from those recorded as suffering from interference as the equipment to perform frequency handoff, assumed to be equipment i. The earliest interference-affected operational phase t is chosen for frequency handoff. The new frequency x_{i2} is formed by perturbing the assigned frequency x_{i1} of equipment i at operational phase t.

The pseudocode for particle variable dimension mutation is shown as follows.

---

01   Let P, P_m denote the mutated particle before and after the mutation, respectively. d denote the dimension.

02   $P_m = P$.

03   Choose an interfered equipment i randomly of P, and record the earliest interfered phase t;

04   $x_{i1} = x_i$;

05   Calculate x_{i2} ;

06   $P_m.x = (P_m.x, x_{i2})$

07   Append $P_m.\text{info}_{\text{dynamic}} = (i, t)$;

08   Calculate $P_m.f_1$;

09   **If** $P_m.f_1 < P.f_1$

10   Mutation success!

11        $P_m.d = P_m.d + 1$;

12   Calculate $P_m.f_2$;

13   **else**

14   Mutation fail.

15   **End if**

DHMSPSO was tested on the FAP. To simplify the problem, each operational platform was configured with 5 deployed frequency-dependent equipment. The selectable frequencies for these equipment range from 20 to 100 MHz. The algorithm parameters of DHMSPSO are set as follows: particle swarm size $n = 10$, external archive size $NA = 10$, maximum number of iterations $T_{max} = 200$, learning factors $c_1 = 1.5$ and $c_2 = 1.5$, maximum inertia weight $\omega_{max} = 0.5$, and minimum inertia weight $\omega_{min} = 0.1$. Other parameters are the same as in the paper.

Depending on the scale of the problem, the number of operating platforms is set to 5, 7, and 10. The simulation results are shown in Fig. 9.19.

As shown in Fig. 9.19, the results indicate that as the number of frequency-dependent equipment increases, spectrum resources become increasingly scarce, and the level of EMI among equipment rises accordingly. Through the allocation and optimization of frequency resources by DHMSPSO, EMI can be minimized efficiently via frequency handoff. Furthermore, by incorporating Pareto optimality theory, the algorithm preserves multiple non-dominated solutions, thereby providing diverse frequency assignment schemes for selection. The control system can weigh the EMI cost and the number of frequency handoff to select and adopt the most suitable frequency assignment scheme.

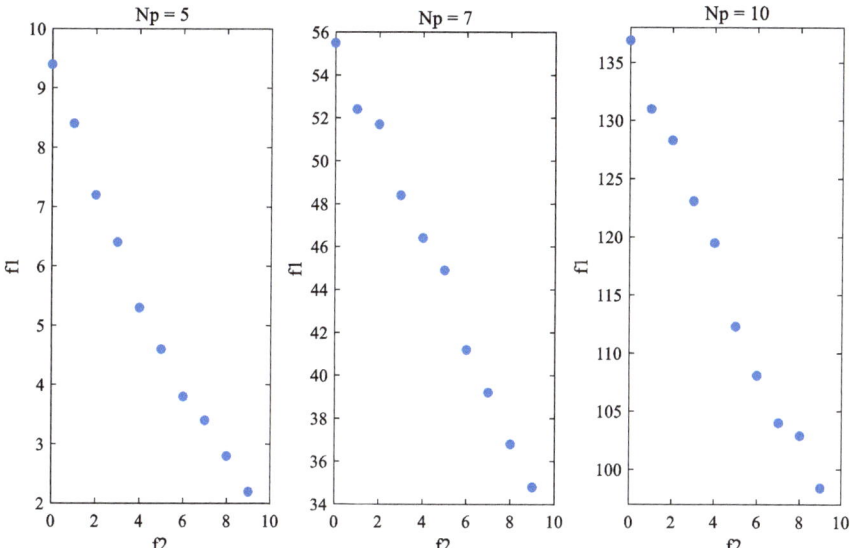

Fig. 9.19 Simulation optimization results of DHMSPSO for FAP

9.6 Summary

This chapter presents evaluation metrics for assessing multi-objective optimization algorithms. It then introduces two widely used MOPSO algorithms (CMOPSO and MOCLPSO) and proposes improvements to them, followed by the design of data experiments to compare their performance. Subsequently, based on the characteristics of MOPSO algorithms, a distance-based PSO improvement algorithm (DISMOPSO) is proposed, with data experiments designed to evaluate its performance. Finally, a novel multi-swarm cooperative MOPSO (CMPSO) is introduced, and its performance is compared with other algorithms.

References

1. Coello CA, Veldhuizen DAV, Lamont GB (2002) Evolutionary algorithms for solving multi-objective problems. Kluwer Academic Publishers
2. Meng HY (2005) Research on multi-objective evolutionary algorithms and their applications [J]. Math Appl
3. Hu X, Shi Y, et al (2004) Recent advances in particle swarm [J]. In: Congress on evolutionary computation, 2004 CEC2004
4. Li X (2004) Better spread and convergence: particle swarm multiobjective optimization using the maximum fitness function [J]. Lect Notes Comp Sci 3102:117–128
5. Cagnina L, Esquivel S, Coello Coello AC (2005) A particle swarm optimizer for multi-objective optimization [J]. J Comput Sci Technol 5(4):204–210

6. Zhao B, Cao Y-J (2005) Multiple objective particle swarm optimization technique for economic load dispatch [J]. Zhejiang Univ Sci 6A(5):420–427
7. Lei DM, Yan XP (2009) Multi-objective intelligent optimization algorithms and their applications[M] [J]. Science Press, Beijing, p 38
8. Zhan ZH, Li JJ, Cao JN (2013) Multiple populations for multiple objectives: a coevolutionary technique for solving multiobjective optimization problems[J]. IEEE Trans Cybernet 43(2):445–463
9. Chen X, Feng X, Jiang X, et al (2025) A Dynamic heterogeneous multi-swarm PSO for multi-objective frequency assignment problem. Exp Syst Appl 289(15):128295